Daniel Reid Kuespert
Research Laboratory Safety
De Gruyter Graduate

Also of interest

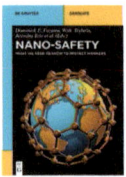
Nano-Safety.
What We Need to Know to Protect Workers
Fazarro, Trybula, Tate, Hanks (Eds), 2016
ISBN 978-3-11-037375-2, e-ISBN 978-3-11-037376-9

Process Technology.
An Introduction
De Haan, 2015
ISBN 978-3-11-033671-9, e-ISBN 978-3-11-033672-6

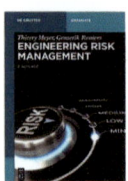
Engineering Risk Management.
3rd Edition
Meyer, Reniers, 2016
ISBN 978-3-11-041803-3, e-ISBN 978-3-11-041804-0

Energetic Materials Encyclopedia.
Klapötke, 2018
ISBN 978-3-11-044139-0, e-ISBN 978-3-11-044292-2

Industrial Inorganic Chemistry.
Benvenuto, 2015
ISBN 978-3-11-033032-8, e-ISBN 978-3-11-033033-5

New Series at De Gruyter:

Integrated Security Sciences
Reniers, Khakzad, van Gelder (Series Eds.)
ISSN 2367-0223, e-ISSN

Daniel Reid Kuespert

Research Laboratory Safety

—

DE GRUYTER

Author
Daniel Reid Kuespert, PhD.
Johns Hopkins University
Krieger School of Arts & Sciences
Whiting School of Engineering
103G Shaffer Hall
3400 North Charles St.
Baltimore MD 21218
dkuespert@jhu.edu

ISBN 978-3-11-044439-1
e-ISBN (PDF) 978-3-11-044443-8
e-ISBN (EPUB) 978-3-11-043705-8

Library of Congress Cataloging-in-Publication Data
A CIP catalog record for this book has been applied for at the Library of Congress.

Bibliographic information published by the Deutsche Nationalbibliothek
The Deutsche Nationalbibliothek lists this publication in the Deutsche Nationalbibliografie; detailed bibliographic data are available on the Internet at http://dnb.dnb.de.

© 2016 Walter de Gruyter GmbH, Berlin/Boston
Typesetting: Konvertus, Haarlem, NL
Printing and binding: CPI books GmbH, Leck
Cover image: kasto80/iStock/Thinkstock
♾ Printed on acid-free paper
Printed in Germany

www.degruyter.com

Contents

Part I: Introductory Material

1 Introduction — 3
1.1 Accidents in the research laboratory — 3
1.1.1 Vladimir Likhonos: eating explosives — 3
1.1.2 Karen Wetterhahn: a deadly droplet — 4
1.1.3 Michele Dufault: hair is a hazard — 4
1.1.4 Louis Slotin: A slipped screwdriver — 6
1.1.5 Preston Brown: Ignoring safety protocols — 7
1.1.6 Sheri Sangji: a spontaneous fire — 9
1.2 Factors contributing to laboratory accidents — 9
1.2.1 Reason's Swiss cheese model — 9
1.2.2 Accident "causes" — 11
1.2.3 Unsafe conditions versus unsafe behavior — 11
1.3 Hazards in the laboratory — 13
1.3.1 Types of hazards — 13
1.3.2 Main risks in laboratories — 14

2 Ethical responsibilities — 15
2.1 Who requires protection? — 15
2.2 Ethical responsibility to others in the lab — 16
2.3 Penalties for ethical violations — 18

3 Assessing and controlling risk — 20
3.1 Distinguishing hazard from risk — 20
3.2 Simple methods for estimating risk — 20
3.3 A semiquantitative method for risk estimation in the laboratory — 22
3.4 Risk assessment exercises — 25

4 Hazard and risk controls — 26
4.1 The hazard control process — 26
4.1.1 Hazard identification — 26
4.1.2 Risk screening — 27
4.1.3 Hazard analysis — 28
4.1.4 Hazard control — 28
4.2 Classifying hazard controls — 29
4.2.1 Functional classification of hazard controls — 29
4.2.2 Traditional hierarchy of controls — 30

4.2.3	Creativity in hazard control —— 31	
4.3	Exercises: Hazard control —— 31	

Part II: Hazard classes and control methods

5	**Hazard identification methods —— 35**	
5.1	Brainstorming, mind-mapping, and other creative methods —— 35	
5.2	Checklists —— 36	
5.3	Reference books —— 36	
5.4	Regulations and standards —— 37	
5.5	Real-life hazard identification —— 38	
5.6	Exercises: Hazard identification —— 39	
6	**Physical hazards —— 40**	
6.1	Mechanical hazards —— 40	
6.1.1	Pinch points —— 40	
6.1.2	Guards and interlocks for mechanical hazards —— 40	
6.1.3	Shear points —— 42	
6.1.4	Run-in points —— 43	
6.1.5	Wrap points —— 45	
6.1.6	Clobbering —— 47	
6.2	Sharps —— 48	
6.2.1	What is a sharp? —— 49	
6.2.2	Sharps handling —— 49	
6.2.3	Sharps disposal —— 50	
6.3	Heat —— 52	
6.3.1	Common laboratory sources of heat —— 52	
6.3.2	Heat-protective apparel —— 53	
6.3.3	Using torches, burners, and other open flames in the lab —— 54	
6.4	Cold (including cryogen safety) —— 55	
6.4.1	Common laboratory sources of low temperatures —— 56	
6.4.2	Safe procedures for maintenance of refrigerators and freezers —— 56	
6.4.3	Cryogenic temperatures —— 57	
6.5	Pressure and vacuum —— 63	
6.5.1	Compressed gases —— 63	
6.6	Electricity and magnetism —— 73	
6.6.1	Electricity —— 74	
6.6.2	Magnetism —— 78	
6.7	General environmental hazards —— 80	
6.7.1	Trips, slips, and-falls —— 80	

6.7.2	Lighting —— 81	
6.7.3	Noise —— 81	
6.7.4	Security hazards —— 81	
6.8	Case study: Chemistry experiment —— 82	
6.9	Exercises: Physical hazards —— 84	

7 Chemical hazards —— 86
7.1	Routes of exposure to chemical hazards —— 86
7.2	Chemical properties contributing to hazard —— 86
7.2.1	Reactivity —— 86
7.2.2	Volatility —— 87
7.3	The chemical fume hood —— 87
7.4	General hazard classifications and precautions —— 90
7.4.1	Experimental protocols for chemical handling —— 90
7.4.2	Flammables and oxidizers —— 91
7.4.3	Corrosives —— 100
7.4.4	Toxics —— 104
7.4.5	Physical hazards from chemicals —— 110
7.4.6	Reactive chemicals —— 110
7.5	Communicating chemical hazards —— 113
7.5.1	NFPA 704 "fire diamond" —— 113
7.5.2	Transportation labeling —— 114
7.5.3	The Globally Harmonized System —— 115
7.5.4	The Safety Data Sheet —— 116
7.6	Case studies —— 118
7.6.1	Chemistry experiment —— 118
7.6.2	Biology experiment —— 121
7.7	Exercises: Chemical hazards —— 124

8 Biological hazards —— 125
8.1	Lab-acquired infections —— 125
8.2	Assessment of biological infection risk —— 126
8.2.1	Agent hazards —— 126
8.2.2	Laboratory procedure hazards —— 128
8.3	Biosafety levels —— 129
8.3.1	Biosafety level 1 (BSL–1) —— 129
8.3.2	Biosafety level 2 (BSL–2) —— 130
8.3.3	Biosafety level 3 (BSL–3) —— 130
8.3.4	Biosafety level 4 (BSL–4) —— 132
8.4	Biological laboratory work practices —— 134
8.4.1	General laboratory practices —— 134
8.4.2	Personal protection —— 135

8.4.3	Pipetting, syringing, and other sample-transfer methods —— 135	
8.4.4	Equipment use —— 136	
8.4.5	Storage, inventory, and labeling —— 136	
8.5	The biological safety cabinet —— 137	
8.5.1	A BSC is not a chemical fume hood —— 139	
8.5.2	The "laminar flow hood" or "clean air hood" is not a BSC —— 139	
8.5.3	Using a BSC —— 141	
8.6	Case studies —— 143	
8.6.1	Biology experiment —— 143	
8.6.2	Civil/environmental engineering experiment —— 145	
8.7	Exercises: biological hazards —— 146	

9 Radiation hazards —— 148
9.1   Ionizing radiation —— 148
9.1.1   Types of ionizing radiation —— 148
9.1.2   Sources of hazard from ionizing radiation —— 148
9.1.3   Control of ionizing radiation —— 151
9.2   Non-ionizing radiation —— 153
9.2.1   Ultraviolet radiation —— 153
9.2.2   Infrared radiation —— 154
9.2.3   Radiofrequency (RF) radiation —— 155
9.2.4   Laser light sources —— 156
9.3   Case studies —— 162
9.3.1   Chemical engineering experiment —— 162
9.3.2   Medical experiment —— 164
9.3.3   Exercises: radiation hazards —— 166

Part III: Hazard analysis techniques

10 The Checklist technique —— 171
10.1   Strengths, weaknesses, and suitability —— 171
10.2   Sources of checklists —— 171
10.3   Example checklist: Quick laboratory inspection —— 172
10.4   Evaluating recommendations from hazard analyses —— 175
10.5   Exercises: laboratory inspection —— 175

11 The Job Hazard Analysis technique (JHA) —— 176
11.1   Strengths, weaknesses, and suitability —— 176
11.2   Technique —— 176
11.3   Example JHA —— 178
11.4   Exercises: Job Hazard Analysis —— 178

12 The What-If? technique —— 180
- 12.1 Strengths and weaknesses —— 180
- 12.2 Suitability —— 180
- 12.3 What-If? Technique —— 181
- 12.3.1 Scoping —— 181
- 12.3.2 Team assembly —— 183
- 12.3.3 What-If? —— 183
- 12.3.4 Causes —— 185
- 12.3.5 Consequences —— 185
- 12.3.6 Controls —— 186
- 12.3.7 Current risk —— 186
- 12.3.8 Recommendations —— 187
- 12.3.9 Revised risk —— 188
- 12.4 Example What-If? Study: Multi-axis press —— 188
- 12.4.1 Nodes —— 191
- 12.4.2 Team assembly —— 191
- 12.4.3 What-if #1: What if a hydraulic actuator fails? —— 191
- 12.4.4 What-if #2: What if a hydraulic line fails? —— 192
- 12.4.5 What-if #3: What if the hydraulic pump develops a leak? —— 193
- 12.5 Exercise: What-if? technique —— 195

Part IV: Practical applications of hazard control

13 Controlling hazards in a laboratory procedure using JHA —— 199
- 13.1 Reproducing a procedure from the literature —— 199
- 13.2 Exercises: Using procedures taken from a research paper —— 201

14 Evaluating risks in an experimental apparatus using What-If? technique —— 202
- 14.1 Case study: What-If? technique —— 202
- 14.2 What-If? technique study on an experimental apparatus —— 206

15 Designing an experiment from scratch —— 207
- 15.1 Hazard controls are *ex post facto* solutions —— 207
- 15.2 The only set factor in an experiment is the objective —— 207
- 15.3 Inherently safer design principles —— 208
- 15.3.1 The history of ISD —— 208
- 15.3.2 ISD design principles —— 209
- 15.4 Case studies in laboratory ISD —— 210
- 15.4.1 Lab ISD case study: Impact testing of steel —— 210

15.4.2	Lab ISD case study: The "Rainbow Experiment" —— **211**	
15.5	Exercise: Inherently safer design of a hazardous experiment —— **212**	

Part V: Appendices

16 **Laboratory safety checklists (abbreviated) —— 215**

17 **Checklist reviews for common laboratory operations —— 223**
17.1 Delivering gas from a compressed gas cylinder —— **223**
17.2 Flame-sealing a glass tube with an oxyacetylene torch —— **226**
17.3 Using a biological safety cabinet —— **228**

18 **Writing experimental protocols and Standard Operating Procedures —— 230**
18.1 Types of "SOP" —— **230**
18.2 General advice on writing protocols —— **230**
18.3 Writing protocols for hazardous materials handling —— **232**
18.4 Writing protocols for experimental procedures —— **239**
18.5 Writing protocols for use of hazardous equipment —— **242**

19 **Annotated bibliography of laboratory safety references —— 247**

References —— 250

Index —— 257

Preface

While research safety is a central and critically-necessary component of successful scientific and engineering research, it is often learned (if at all) by "osmosis," that is, by placing the student or young professional in a laboratory environment and expecting him or her to divine appropriate methods for keeping him- or herself and his or her staff safe. If the researcher is lucky, a number of reference books such as **Prudent Practices in the Laboratory** are provided, along with some sketchy Standard Operating Procedures and half-hearted on-the-job training ("Don't stick your fingers in here; they'll get caught.")

This book is an attempt to address that situation. It is an introduction to the range of safety hazards encountered in the laboratory and some practical methods that may be used to identify and control those hazards. The book is designed for readers with limited experience in an independent research environment (graduate students or early-stage professionals) but with at least some experience in laboratory classes or teaching laboratories. It will also be useful to non-laboratory middle-management newly assigned to supervise laboratories.

A successful researcher must learn the limits of his or her job—that is, when information should be sought from others and when one is expected to either look it up or discover it for oneself experimentally. This book is written with that in mind. It is not intended to be a specialist reference, but extensive citations are provided to assist the researcher in finding more information. The Appendix entitled **Annotated bibliography of laboratory safety references** (see chapter 19 below) provides references to more specialized works that would be especially suitable in the researcher's library. In some cases, the reader is referred to qualified specialist personnel, such as health physicists, biosafety professionals, or occupational health physicians.

The choice of material herein is quite deliberate. For example, an extensive chapter is devoted to physical hazards in the laboratory; this is partially because of the breadth of physical hazards to be found in a lab. It is also a reflection of the fact that most reference books on laboratory hazards tend to be oriented principally toward chemical, radiation, or biological safety, and they mention underlying physical hazards only in passing. Because ionizing radiation is usually subject to comprehensive regulation and control, less information has been provided—in most areas handling radioisotopes and other ionizing radiation sources, a health physicist will be available to provide training and advice.

The author and publisher have made best efforts to ensure that the information in this book is as up-to-date and accurate as possible. Nevertheless, no warranty is made as to the accuracy or precision of that information, because safety knowledge can and does change—sometimes quickly. The reader is urged to identify sources of information on current safety events in his or her field (an example being **Chemical**

& Engineering News' frequent features on dangerous chemical reactions) to stay current on the latest safety understanding.

One final note: this book does not address first aid and medical treatment except in passing. The reader is referred to the local chapter of the Red Cross or Red Crescent for certified first-aid training.

Acknowledgements

Acknowledgements to:
- My wife Martha & daughter Elisabeth, for their support while researching and writing this book.
- Dr. Mark Benvenuto, for convincing me to write this text.

My readers:
- Mr. Nathaniel (Niel) Leon, mechanical design engineer, Laser Safety Advocate and staff engineer for Johns Hopkins University (JHU) Engineering Design Services.
- Mr. Stanley Wadsworth, Radiation Safety Officer, Johns Hopkins University Department of Health, Safety, and Environment.

My case studies and examples:
- Dean Beverly Wendland (JHU Krieger School of Arts & Sciences) and Mr. Kyle Hoban.
- Dr. Tyrel McQueen (JHU Chemistry) and Mr. Zachary Kelly.
- Dr. Rebekka Klausen (JHU Chemistry) and Ms. Heidi van der Wouw.
- Dr. Edward J. Bouwer (JHU Geography & Environmental Engineering).
- Dr. Todd Hufnagel (JHU Materials Science & Engineering) & Dr. Ravi Shivariman.
- Dr. Kenneth Karlin (JHU Chemistry), Dr Savita Sharma, and Messrs. Jeffrey Liu and David Quist.
- Dr. David Gracias (JHU Chemical & Biomolecular Engineering) and Dr. Shalaka Dewan.
- Dr. Bernhard Fuerst (JHU Laboratory for Computational Sensing and Robotics) and Mr. Sing Chun Lee.
- Dr. K.T. Ramesh (JHU Hopkins Extreme Materials Institute) and Mr. Matt Shaeffer.
- Dr. Patricia Janak (JHU Psych & Brain Sciences, JHMI Neuroscience) and Dr. Flavia Barbano.

Photographic assistance:
- Mr. Brian Crawford (JHU Materials Science & Engineering).
- Mr. Boris Steinberg (JHU Chemistry).
- Mr. Perry Cooper (JHU Health, Safety, and Environment).
- Mr. Richard Middlestadt (JHU Engineering Manufacturing).

Special sources:
- Ms. Yvonne Morris, Director, Titan Missile Museum, Titan Missile National Historic Landmark, for an original copy of the Titan II launch checklist.
- Ms. Jillian Emerson, U. California (Davis) Department of Chemistry, for permission to borrow from her protocol for a safe version of the "Rainbow Experiment."

- Mr. Richard Malenfant, Los Alamos National Laboratory (ret.), for fruitful discussions about the death of Louis Slotin.
- Dr. Ben Shafer (JHU Civil Engineering) for essential information about hydraulic presses.

Note: Unless otherwise credited, all photographs are by the author.

Notes to the Instructor

This book is primarily intended as a textbook as opposed to a reference book, although the table of contents and indices have been constructed to permit the book's users to locate topical information for reference as well.

The book is designed to be used by individual practitioners, particularly at the early-stages of their career; by students, particularly first-year graduate students and advanced undergraduates; and by those who supervise laboratories. Exercises and text presume familiarity with a broad range of disciplines such as would be developed by a well-educated scientist or engineer.

Additional instructor's notes, including answers (or possible answers for open-ended questions) to exercises, are provided online at http://www.degruyter.com/view/product/462533?rskey=SNfwfG

The overall structure of the text is to follow the common method of hazard control:
1. Identify the hazards of the system, equipment, or procedure;
2. Assess the risk of each hazard to identify those requiring further analysis and control;
3. Analyze each of the major hazards to identify possible modes of exposure, causes, consequences, and appropriate controls;
4. Evaluate the proposed controls based on criteria of effectiveness, total risk reduction achieved, and cost;
5. Implement the controls;
6. Monitor the system for changes or new information that may affect the hazard analysis.

The chapters on hazard identification follow the four basic types of hazard found in the laboratory: physical, chemical, biological, and radiation. Usually, each chapter is capped by two brief case studies based on actual research taking place at the author's institution—one case study is for the "typical" discipline experiencing that type of hazard (e.g., chemistry usually faces chemical hazards). The second case study points out the "unusual" hazards of another discipline—the physical hazards that occur in chemistry laboratories, for example.

The exercises at the end of each chapter are meant to be adaptable to your particular needs. Usually there are one or more simple matching or multiple-choice questions as preparation, followed by a series of situationally-based activities based on different disciplines. These open-ended questions allow the student to apply the techniques described in the chapter—identifying physical hazards, for example. Feel free to substitute your own laboratory situations and research occurring at your institution for these scenarios. For those using the provided scenarios, simple, moderate, and advanced lessons learned are provided online as samples of results to be expected of students at various levels.

The last sections of the book, save for the Appendices, concentrate on application of hazard analysis to real (or close-to-real) experimental research protocols, apparatus, and other laboratory situations. The final chapter discusses design of experiments—and the principles of Inherently Safer Design (ISD) [1], a chemical engineering technique easily adaptable to laboratory situations. ISD removes the need for explicit hazard control by attempting to eliminate or reduce risk at design time—substituting a safer chemical solvent or miniaturizing a pressure vessel, for example.

Part I: **Introductory Material**

1 Introduction

Science and engineering research concerns **new knowledge, new technology**. The scientific method depends on a hypothesis, a question with an unknown answer. The research environment, therefore, necessarily involves operations and materials whose behavior and properties are not fully understood. Lack of understanding increases the risk of working in a research laboratory or field location, and practicing scientists and engineering researchers must often take steps to control that risk and reduce it to a tolerable level.

1.1 Accidents in the research laboratory

Accidents in research laboratories are common; they may involve injury, illness, damage to equipment, or simple failure of the experiment. Firm data on the incidence of laboratory accidents is not available, partially because health authorities typically do not identify incidents as lab accidents **per se**, and partially because employers sometimes keep incidents confidential (to avoid reputational losses or lawsuits). Nevertheless, the most serious incidents, those which require the involvement of civil authorities such as firefighters or occupational safety investigators, become widely known. Six notorious examples are summarized below, four in chemistry, one in physics, and one which involved machine shop equipment. Each example includes both an obvious lesson to be learned and a more subtle one. In all but one case, the investigator died.

1.1.1 Vladimir Likhonos: eating explosives

In early December 2009, Vladimir Likhonos, a chemistry student at the Kiev Polytechnic Institute in Ukraine, was working at home on his computer. Mr Likhonos enjoyed chewing gum while working, and he often dipped his gum in a small dish of citric acid that he kept on his desk (to give it a tangy flavor).

Apparently, at some point in the recent past, he had brought a small container of an explosive chemical home from the laboratory for an unknown reason. It is theorized that while concentrating on his work, he absent-mindedly dipped his gum in the explosive instead of in the citric acid. When he bit into the explosive-laced chewing gum, the resulting explosion blew off his jaw and severely injured his face; he consequently bled to death [2].

The simple lesson to be drawn from this incident is that food and drink should **never** be stored or consumed near hazardous chemicals. A more general lesson might be that laboratory operations and products should always be kept strictly separate from day-to-day activities: do not remove lab coats from the laboratory, nor wear gloves outside the lab, for example.

1.1.2 Karen Wetterhahn: a deadly droplet

On 14 August 1996, Dr Karen Wetterhahn, professor of chemistry at Dartmouth College (Hanover NH, US), was conducting ^{199}Hg nuclear magnetic resonance (NMR) experiments [3]. Wetterhahn, 48, was an expert on heavy-metal toxicity—she was well-known for elucidating the mechanism of carcinogenicity for Cr^{6+} [4]. NMR spectroscopy requires use of a reference compound, as the spectrum of the sample under analysis is expressed relative to the reference peak. At the time, the standard ^{199}Hg NMR reference material was dimethylmercury [5]. This compound is known to be extremely hazardous (all known exposures have been fatal [3]), and Wetterhahn handled it only in a chemical fume hood while wearing disposable latex gloves [6].

While transferring dimethylmercury to a NMR sample vial, Wetterhahn spilled one or two drops of the compound on her latex-gloved hands. Dimethylmercury can penetrate disposable latex gloves in under 15 seconds [7]. Since she was not aware of this fact, Wetterhahn tidied up the spill and removed the gloves at the end of the operation.

In January 1997, Wetterhahn was admitted to the hospital with a diagnosis of severe mercury poisoning, and despite aggressive chelation therapy and other treatments, she died 8 June 1997.

The clear lesson to be learned from Wetterhahn's death is that when handling compounds known or suspected to be highly toxic, it is essential to **fully** understand the properties of the material and to take additional precautions, such as using multiple layers of different gloves to slow penetration and changing gloves at the first sign of exposure. In this case, the "supertoxicity" of dimethylmercury had been known since its first synthesis in 1863, within months of which both lab technicians who had performed the procedure died horribly [8].

Another lesson is more subtle: the manner in which Wetterhahn conducted her operation created a situation where her gloves were the **only** line of protection between her skin and the compound. **Personal protective equipment** (universally known as **PPE**) is notoriously unreliable. **Never** rely on PPE as the only line of defense against an incident unless absolutely necessary, and if it appears to be absolutely necessary to do so, obtain professional advice on selection and use of the equipment.

1.1.3 Michele Dufault: hair is a hazard

A particularly sad incident occurred on 12 April 2011, at Yale University. Physics and astronomy student Michele Dufault was working late at night in a machine shop. She was working on her senior thesis: a device employing liquid helium for detection of dark matter in the universe [9].

Ms. Dufault was using a lathe, a machine tool used for producing axially-symmetric pieces of materials such as wood or metal. The workpiece is clamped with a pair of chucks to hold it steady, and it is spun at great speed. Sharp tools placed against the workpiece quickly cut any shape desired, provided that it has a circular cross-section.

Figure 1.1: Machinist Richard Middlestadt uses a lathe to shape a part.

Her hair was suddenly caught in the lathe, which would have quickly wound it in like thread on a spool. This is an example of a **wrap point** (see section 6.1.5 below). The shop was unmanned outside business hours, and she died of asphyxia before she was discovered by other students working in the building.

A medical examiner ruled her death an accident caused by neck compression [9]. Michele Dufault was a trained and qualified individual, by the standards of her institution. Machine shop rules and training materials specifically warned of putting hair up in shop to prevent injury: "…if your hair is caught in spinning machinery, it will be pulled out if you are lucky." Dufault was **not** lucky in this instance.

A simple lesson that can be drawn from Michele Dufault's death is that one should **never work alone** when performing work posing significant risk. It was well-known that lathes can catch hair and cause serious injury [10]. The presence of a "buddy" in the shop during hazardous operations would have provided someone to shut down the machine, render first aid, possibly extract Ms. Dufault from the lathe, and call for help.

A more subtle lesson is that oftentimes in the rush to complete a research project, it is very easy to fall prey to a very simple hazard, even if one has been trained and warned of its presence. Training and warning signs are examples of **administrative controls** (see section 4.2.2 below), and one reason other, more engineering-based controls (such as guards) are preferred over administrative controls is that they are more reliable. (Note that it is not possible to place a guard over the spinning workpiece in a standard lathe, though, so Yale and Michele Dufault were forced to fall back on administrative controls.)

1.1.4 Louis Slotin: A slipped screwdriver

On 30 May 1946, Dr. Louis Slotin, a physicist at what is now the US Los Alamos National Laboratory, which designed and prototyped the first atomic bombs, conducted an experiment to determine criticality thresholds in a plutonium core assembly. The experiment involved holding two Pu spheres slightly apart with a screwdriver and measuring neutron emission as they were brought closer and closer [11]. Slotin had performed this procedure several times; he was demonstrating it to a new researcher who was to take over the experiment, but he was in somewhat of a hurry to conduct the demonstration ([12]; as duplicated in [13]).

The screwdriver slipped, and the plutonium spheres were brought into full contact; the safety wedges provided to prevent such contact also slipped. The presence of Slotin's hand in a hole as he held the upper sphere inadvertently boosted the reactivity of the assembly (the high water content of flesh makes it an excellent neutron moderator, thereby slowing some of the neutron emissions already present and boosting the reactivity of the assembly). This created a supercritical assembly from which neutron emission immediately increased drastically [14]. There was a blue flash as the surrounding air was ionized by the strong radiation, and Slotin immediately disassembled the experiment by flipping off part of the beryllium neutron reflector surrounding the plutonium, possibly preventing an actual blast from additional energy release.

Slotin was exposed to approximately 21 Sv (2100 rem) of neutron radiation [11], and others in the room were exposed to lesser amounts. The amount of neutron radiation required for lethality varies with the energy of the neutrons, but Slotin's exposure was orders of magnitude over fatal levels—in the US, 750 mSv or so (10 mSv times your age) is considered a maximum **lifetime** dose [15]. Slotin died 9 days later from acute radiation syndrome [11].

Figure 1.2: Recreation of Louis Slotin radiation incident. Los Alamos National Laboratory Historical Photograph (public domain).

One lesson from this incident is quite simple: do not hurry hazardous experiments; as Raemer Schreiber, a physicist present at the accident, put it, "New critical assemblies should never be reduced to a routine matter to be 'run through before lunch'" [13]. More general, though, is the lesson not to become blasé about experiments that are known to be dangerous, as this one was. Indeed, Enrico Fermi, upon being told of the experimental technique in use some time earlier, remarked to Slotin, "Keep doing that experiment that way and you'll be dead within a year" [16]. Schreiber in his memo to Los Alamos authorities reporting the incident remarked, "As soon as a person ceases to be nervous about the work he should be transferred to another job."

1.1.5 Preston Brown: Ignoring safety protocols

On 7 January 2010, graduate student Preston Brown, a chemistry student at Texas Tech University required several grams of nickel hydrazine perchlorate, an "energetic material" (a synonym for "explosive compound") under study by his laboratory. Despite (verbal) lab protocols specifying that batches of such compounds be limited to 100 mg each, Brown and a colleague began a synthesis of **10 g** of the chemical in a single batch, apparently to reduce errors from batch-to-batch inconsistency [17].

The product had a lumpy texture, and Brown transferred half of the batch to a mortar-and-pestle. He proceeded to **grind** the compound (compounds of this type are known to be shock-sensitive), using no blast shield, believing it to be safe because the compound was wet. After finishing, he removed his safety goggles, but he returned shortly thereafter to stir the material an additional time. He did not replace his safety equipment before doing so. The perchlorate detonated [17].

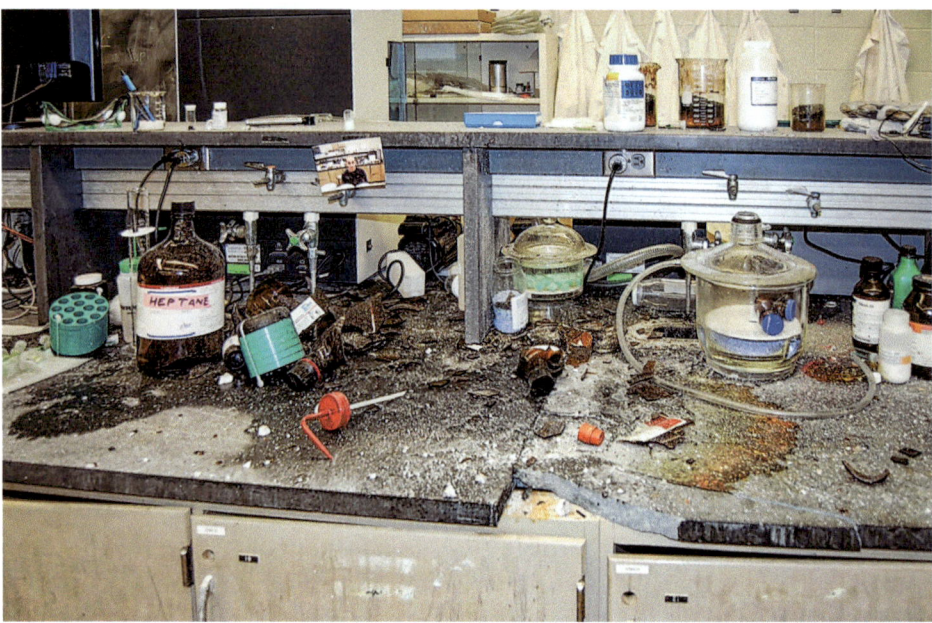

Figure 1.3: Texas Tech University laboratory explosion, 2010. Photo courtesy Texas Tech University.

Brown "lost three fingers on his left hand, burned his hands, and injured his eyes" [18]. The explosion blew a hole in the chemical-resistant resin bench top and injured a second researcher as well.

The obvious lesson learned from this incident is that materials having known potential for explosive behavior must be treated with extreme care. A key, not-so-obvious lesson is that, at least for high-risk procedures, experimental protocols and operating limits should be **in writing, with all persons who are to use them adequately trained in their use.**

1.1.6 Sheri Sangji: a spontaneous fire

The most notorious laboratory incident of recent times is the death of Sheri Sangji, laboratory assistant in an organic chemistry lab at the University of California at Los Angeles on 29 December 2008. Sangji was attempting to transfer a solution of **tert**-butyllithium (abbreviated **tBuLi** by chemists) in a highly flammable solvent (pentane) from a sealed reagent bottle using a plastic syringe too small for the task. Typically, a syringe should never be filled more than 50% full when transferring hazardous materials. Sangji withdrew 53 mL of solution into a 60 mL syringe, using a very short needle (also not a recommended practice) [19, 20].

Sangji accidentally pulled the syringe plunger out of the barrel, dousing herself with the contents of the reagent bottle, which she had pressurized with nitrogen to assist the transfer. **tBuLi** is **pyrophoric**, that is, it spontaneously combusts in air, and she was immediately set ablaze. She had no lab coat on and was wearing a highly combustible acrylic sweater. Despite the safety shower immediately behind her, neither she nor any of her co-workers activated the shower to put out the fire and remove the **tBuLi**. She died 18 days later from her burns [20, 21].

Obviously, using appropriate techniques to transfer pyrophoric reagents [19] is essential when using them. The real lesson in this case, though, is that laboratory managers and principal investigators must take responsibility for overseeing safety in their laboratories. Sangji was not trained in use of pyrophoric materials, was not provided with a laboratory coat (fire-retardant or otherwise), and was operating in a laboratory where safety information was typically communicated verbally, if at all [20, 21]. All of these factors are the responsibility of the individual operating the laboratory—indeed, the principal investigator, Dr. Patrick Harran, was charged with several criminal violations of occupational safety laws. In a legal agreement struck to avoid a trial, Dr. Harran agreed, among other things, not to deny responsibility for the laboratory conditions on the day of Sangji's death [22].

1.2 Factors contributing to laboratory accidents

Many factors contribute to laboratory accidents, and there is rarely a single "cause" that can be definitively identified. In most cases, an incident is the result of multiple causes and conditions coming together to produce an undesired effect.

1.2.1 Reason's Swiss cheese model

One simple model of accidents was developed by Reason [23], whimsically titled the "Swiss cheese model." Reason models a single "cause," which is typically blocked from actually creating an incident by a series of layers of protection. No protective measure is 100% effective, and often there are several "holes" in the layer, much like a slice of Swiss cheese.

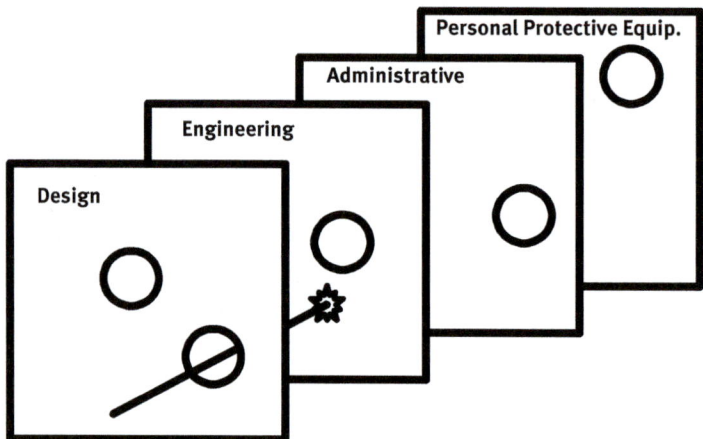

Figure 1.4: Reason's "Swiss Cheese" model of accidents.

Safety eyewear is an example: safety glasses must be properly fitted, chosen for the application, and actually worn by the researcher to be effective. Even so, a large fraction (possibly as much as 35%, depending on the source consulted) of eye incidents occur to persons who are wearing the correct eyewear. If the "holes" in multiple layers align, this creates a vulnerability through which a particular cause can create an incident.

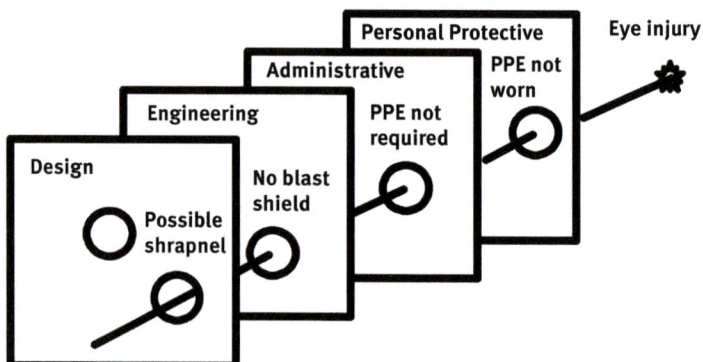

Figure 1.5: Reason's "Swiss Cheese" model of accidents with convergent vulnerabilities.

1.2.2 Accident "causes"

In incident analysis, we distinguish between **proximate causes** of an incident and the **root causes**. Proximate causes are those factors which are immediately responsible for the incident: a flying bolt striking a researcher's eye, for example. A **root cause** is a factor which is ultimately responsible for the existence of the proximate cause.

Since an incident is really a chain of events, how far up the chain one travels in search of root causes depends on available information and the focus of the incident investigation. When one is conducting a shallow investigation to explain and prevent further occurrences of similar incidents, the root cause might be "failure (of the researcher) to wear protective eyewear;" if the focus of the investigation is to prevent a wider class of incidents, perhaps "failure (of the principal investigator) to manage and provide protective eyewear" might be considered the root cause. In general, when looking for "root causes," the general desire is to travel as far back as possible and concentrate on management issues as opposed to issues with individual researchers—this is because retraining a single researcher to wear safety eyewear protects that researcher, while ensuring that the entire laboratory is properly supplied and trained protects the entire lab population.

1.2.3 Unsafe conditions versus unsafe behavior

Unsafe conditions, such as a malfunctioning piece of equipment or an unsecured compressed gas cylinder, are thought to be responsible for about 20–25% of incidents in the workplace. This number is contested [24], but there is little argument that **physical conditions** result primarily from **behavioral conditions**.

The author's institution conducts not only physical inspections of laboratory condition, but also **behavioral inspections**, searching for signs that behavior issues may be developing in a particular lab. Since it is difficult to observe behavior directly, both because of the time investment (and consequent investment in human resources) required and because people tend to change behavior when they know they are being observed, a short checklist of "physical indicators of behavior" is used instead. Example questions on this checklist include:
- **Are personnel in the laboratory wearing appropriate personal protective equipment for the hazards to which they are exposed?** If the answer is no, this is a physically-obvious signal that hazards are insufficiently respected by the researchers.
- **Are the laboratory waste cans filled to overflowing?** If the answer is yes, it indicates that researchers are paying care only to their individual tasks and not to the common environment in which they work (a phenomenon political scientists call "the tragedy of the commons").

- **Are hazardous chemicals properly stored when not in use?** Solvent bottles left on benches for days clearly indicate that researchers are too concentrated on their tasks to properly store supplies at the end of the workday. Such laboratories usually have tools and other materials spread everywhere on the benches, since the lab workers immediately drop things they no longer need in the first convenient place and forget about their presence.

Many factors lead to unsafe behavior, including:
- **Inattentiveness**. Particularly in labs with poor housekeeping, not paying attention to the task at hand can be a direct cause of an accident. In any potentially-dangerous situation, it is essential to maintain one's attention at all times. Lack of attention leads to **inattentional blindness**, a phenomenon in which a person does not see things to which he or she is not paying attention. Inattentiveness can come from many sources—tiredness or lack of sleep, emotional disturbances (such as the classic "fight with spouse or close friend"), lack of commitment to the task, or even something as mundane as a poor breakfast. It is essential to monitor one's behavior and mental condition in a hazard-rich environment such as a laboratory to avoid this situation.
- **Tunnel vision.** When working on a familiar task requiring close attention, such as a scientific experiment, one can accidentally attend more to one's **expectation** of the coming sequence of events than to the events themselves. The ability to detect and react to unexpected events—signs that a chemical reaction may be overly vigorous, for example, or the hair-standing-on-end sensation that indicates the presence of high voltages may be lost. This tunnel vision is another form of inattentional blindness.
- **Assumed invulnerability.** Younger researchers (generally those less than 25–30 years old) have not yet recovered from the tendency of adolescents to believe themselves immune from danger. This is partially because of a lack of life experience and partially because adolescent brain development actually continues into the twenties. This psychological bias can lead such researchers to undertake experiments and procedures that carry more risk than would a more experienced scientist or engineer. One of the more important lessons of early adulthood is that one is "only immortal for a limited time" [25].
- **"Checklist zombies."** Frequently in safety practice (as in many other fields), we use checklists to guide complicated procedures, ensuring that the proper steps are done in the proper order, or that a minimal set of hazards is considered before undertaking an experiment. (See chapter 10 below.) The use of checklists is a double-edged sword, as it discourages the user from looking critically at the situation. A so-called "checklist zombie" is an individual who thoughtlessly checks the boxes on a checklist, "pencil-whipping" it instead of supplementing the list with careful thought in order to identify other possible hazards not on the list.

- **Lack of experience with the technique in use.** Students in particular, and researchers in general, often execute laboratory procedures with which they are unfamiliar. In the case of students, this is a necessary consequence of the learning process—to learn a procedure, we must perform it for the first time. In research, the procedure is often novel, something **no one** has done before, and naturally a new technique must be tried for the first time. In both of these situations, incomplete knowledge of the hazards of the operation and the ways in which it can deviate from the expected sequence of events create the risk of an accident.
- **Working alone or after-hours.** Lone work or work conducted after normal business hours (whatever those hours are for the organization) exposes the researcher to **increased accident severity,** as resources that would normally be available—for example, someone to call for medical assistance in the event of a disabling accident—are not accessible. It is sometimes necessary to work alone or late in scientific and engineering research, because experiments may take more than one work shift to complete, and certain tasks must be carried out at certain times. Nevertheless, lone or late work is undertaken much more often than actually necessary, particularly in the academic environment, and when it must be done, additional risk-reduction measures are often called for.

1.3 Hazards in the laboratory

1.3.1 Types of hazards

Laboratories contain several basic types of hazards:
- **Chemical hazards** are any hazard arising from chemical compounds or reactions. These can range from simple flammability (ability to catch on fire easily) to unusual forms of toxicity such as mutagens, which have the ability to damage DNA molecules **in vivo**.
- **Biological hazards** are those arising from living organisms. The term is loosely interpreted—"biohazards" include such things as viruses (which some do not consider living in any normal sense) and biotoxins (highly toxic chemicals that happen to be natural products, e.g., the toxic compound ricin, which is derived from castor beans).
- **Radiation hazards** arise from energy transmitted through space. Examples would be X- and gamma-rays, alpha particles, or laser light.
- **Physical hazards** is a catch-all term encompassing any hazard that involves direct contact with some form of energy—if something is not considered a chemical, biological, or radiation hazard, it is a physical hazard. Examples vary widely, from heat to cold to explosions. Note that there is some overlap regarding what a physical hazard is. Explosive properties of chemicals are "physical hazards," and heat transferred by radiation is sometimes thought as a physical hazard as well.

1.3.2 Main risks in laboratories

Data is not kept on the details of each laboratory accident, but in the author's opinion, the "most risky" items used in laboratories are **hazardous chemicals, compressed gases,** and **lasers.** (See chapter 3 for discussion of the distinction between **hazard** and **risk**). This selection is mainly due to ubiquity—each of these items appear across a broad cross-section of laboratories. A nominally **biological** laboratory will certainly handle hazardous chemicals (e.g., ethidium bromide for gel electrophoresis) as well as compressed gases (e.g., carbon dioxide for controlled-atmosphere incubators) and possibly lasers (e.g., for confocal laser-scanning microscopy). Many of these items are potentially lethal, as well as carrying the risk of a myriad of lesser accidents.

2 Ethical responsibilities

Safety in the lab is not merely a matter of compliance with safety rules and regulations. It is an **ethical issue**: What behavior is morally appropriate in a laboratory situation? Killing or injuring other human beings, intentionally or through negligence, is not a feature of most moral codes (except perhaps those of sociopaths). Appropriate, ethical behavior, therefore, is to take actions that promote safety and avoid taking actions which might result in death or injury to persons.

Professional ethics also specifically includes prohibitions against causing injury—the Hippocratic oath of physicians being the best-known example. Most professions, including science and engineering, have ethical principles agreed upon by members of the profession; these are often formalized in codes of ethics. All engineering fields and most major scientific disciplines of which the author is aware have such codes, and they generally emphasize variations on the theme of "do no harm:"

- "Chemical engineers shall...hold paramount the safety, health, and welfare of the public and protect the environment in the course of their professional duties" [26].
- [Fundamental Canon 1:] "Engineers shall hold paramount the safety, health, and welfare of the public in the performance of their professional duties" [27].
- "Members...have a duty to minimise adverse effects on Health and Safety and to recommend the use of best health and safety practice and give appropriate advice" [28].

In the context of professional ethics codes, the term **"the public"** means, in essence, "everyone except your employer," including customers, co-workers, and arguably oneself. In many countries, scientists and engineers may be held legally liable, either civilly or criminally, for violating these ethical codes, particularly if one possesses an official engineering license.

2.1 Who requires protection?

The variety of populations threatened by laboratory accidents is larger than one would think. Ethical obligations to work safely protect:
- **You.** You have an obligation to protect yourself in most cases. The average person, if asked "Is it wrong to act in a manner that poses a substantial risk to oneself?" will generally answer "yes." More thoughtful subjects might debate "edge cases" such as taking risks to protect another (e.g., saving a child from a speeding bus) or deliberately sacrificing oneself (e.g., if one has terminal-stage cancer), but such situations do not normally arise in laboratories.
- **Your co-workers.** It is generally considered unethical to impose risks on another that you would not take yourself. In fact, it has been observed that people in general will react with hostility to those who force risk upon them—observe the public demonstrations that have occurred around the world in response to non-transparent decisions on chemical and nuclear

plant sites, which chemicals are allowed in the food supply, etc. Many situations in the lab impose risk not only on oneself but on one's co-workers—the case of Alain Louati (see section 2.3, below) is an illustrative example.
- **Personal associates.** If you are injured or killed in the laboratory (or imprisoned for causing injury or death to another), your family will be denied your company, your guidance, and a source of income, among other things. It is also possible to bring biological, chemical, or radiological hazards home, either intentionally or inadvertently through negligence, and cause harm to your family, room-mates, etc.
- **The "public."** As mentioned above, codes of ethics usually interpret "the public" to mean "everyone but you and your employer." Failure to adequately protect the public is a serious ethical lapse and may be "rewarded" with fines and criminal charges. In at least one case of which the author is aware, an engineer was murdered for negligence that resulted in a public disaster [29].
- **The environment.** Air, water, and other natural resources are collective property, essential to the survival of Earth's ecosystem and humanity in particular. Taking action which contaminates the environment, or not taking action to protect it, threatens us all. The author has occasionally noted chemical waste bottles left open in chemical fume hoods, allowing the contents to vaporize and be emitted out the building fume hood stack—where all the local residents must breathe them.
- **Science and engineering.** Unsafe behavior and conditions threaten the very ability of scientists and engineers to continue their work. Science and engineering do not occur **in vacuo**, but in the context of public funding and permission to operate—they occur at public sufferance. Engineers and scientists are investigators and problem-solvers by their very nature. "The public" is the source of the scientific and engineering problems they solve and the funds with which they solve them. The public is not likely to permit science and engineering research to continue if they believe it will create a threat to them.

2.2 Ethical responsibility to others in the lab

One's ethical responsibilities in the laboratory are similar to those in any other workplace. Indeed, although the job titles differ, a rough equivalence between the typical research laboratory "rank order" and that of the industrial world can be easily established. (See Table 2.1.)

It is common to hear that the laboratory—academic or not—is a qualitatively different place from other, more industrially-oriented workplaces such as manufacturing facilities and construction sites. As a result, the argument is often made that "industrial" safety rules and regulations, and even ethical principles, should not apply to laboratories; this is usually coupled with the argument that the principal investigator should be king or queen of his/her domain, able to set and responsible for setting safety standards solely on the basis of his or her specialist knowledge and judgment. **Every single industry with which the author has worked—research, foods, chemicals, medicine, industrial manufacturing, even nonprofit management—has made the same specious argument.**

Differences in mode of work between industries (and research is an industry, even if it occurs under the auspices of an academic institution) can very properly dictate

different approaches to safety, but the objectives remains the same: **do no harm, and do not allow harm to occur through inaction.**

General ethical responsibilities vary by rank:
- **Institution or employer:** Responsible for the collective safety of the laboratories under its auspices (usually a legal obligation as well as an ethical one). Institutions often delegate this responsibility to the Health, Safety, and Environment department, but delegation does not absolve the institution as an entity from its responsibilities—it merely enables implementation of them [30].
- **Principal investigator:** Responsible for ensuring that activity in his/her laboratory (including his/her own activity) occurs in a safe manner. This means that researchers recognize potential hazards and pathways for exposure (hazard scenarios), evaluate the risks from each, fully analyze potentially unacceptable risks, and determine appropriate control measures that reduce those risks to a tolerable level.
- **Senior scientist or engineer/postdoctoral fellow:** Responsible for the safety of his or her own work, consulting with and reporting to the principal investigator as appropriate. Principal investigators often spend little time physically in the laboratory; therefore, the senior-ranked researchers are usually the individuals most available for observation by the lower-ranked. As the graduate students and junior scientists/engineers frequently model the behavior exhibited by those above them, this means that those higher-ranked individuals have an additional ethical responsibility to teach appropriate safety behaviors (e.g., use of appropriate personal protective equipment such as safety eyewear) by their own example.
- **Junior scientist/engineer/graduate student:** Responsible for the safety of his or her own work, under the supervision of the principal investigator and those senior to him/her in the laboratory. Because the junior scientist/engineer or graduate student has less experience in the laboratory's equipment and specialist techniques, he or she has an additional responsibility to ask for help before conducting work he/she does not fully understand.
- **Undergraduate students/technicians:** In most laboratories, technicians and undergraduate students are "extra pairs of hands," given specific directives or (with more capable individuals) delegated a small portion of the overall research. A higher degree of supervision should be exercised by higher-ranking researchers.

Some specific ethical practices should be universal to the laboratory (and of the workplace in general):
- **Place safety above productivity or results.** The principal ethical responsibility of any scientist or engineer is to oneself and "the public," meaning anyone who is not the institution or employer. The "institution" or "employer" in this case is to be interpreted as the impersonal corporate entity—any scientist or engineer has an ethical responsibility not to personally endanger supervisors, managers, institutional staff, etc.

- **Ensure that your own work is conducted safely.** Using unsafe practices or permitting unsafe conditions to develop endangers not only yourself but also others. Further, it encourages others to exhibit the same behavior, degrading the safety culture in the entire laboratory and creating an ongoing increase in risk.
- **Supervise the safety behaviors of those working under you.** Any accident that occurs "on your watch" to people under your direction is your responsibility, and in certain cases failure to exercise due diligence can land you in court or jail. (See section 1.1.6, above.)
- **Actively search for safety hazards and risks.** Do not assume that no hazards exist unless you have looked for them. For example, read Safety Data Sheets (see section 7.5.4 below) before working with chemicals, and evaluate equipment for mechanical hazards before use (see section 6.1 below).
- **Remain aware of the safety implications of other's work.** Be aware of experiments and equipment operating in your environment, including neighboring laboratories if possible. Ask questions of your colleagues about the safety risks of their work.
- **Exhibit appropriate safety behaviors for the benefit of others.** This benefits you as well through enhancing the safety culture of the laboratory.
- **Intervene or consult with higher ranks if necessary to prevent or correct unsafe behavior and conditions.** Allowing an unsafe condition or practice to continue makes you in part responsible for the resultant safety risks—if an incident occurs, it will be **your fault** as much as that of the researcher whose experiment caused the incident. Always consult with the researcher to correct such conditions, if necessary reporting them to the principal investigator or even to institutional representatives or civil authorities. (This latter practice is called **whistleblowing**, and while uncomfortable and even having potentially negative career consequences, it is both an ethical and often a legal imperative [31].)

Table 2.1: Approximate academic research and corporate rank-equivalents

Research rank/level	Corporate equivalent
Institution or employer	Corporation
Principal investigator/senior scientist	Manager
Staff scientist/engineer/postdoctoral fellow	Supervisor/Group leader
Junior scientist/engineer/graduate student	Foreman/master tradesman
Undergraduate student/technician	Line worker/journeyman
Precollege (high school) student	Intern/apprentice

2.3 Penalties for ethical violations

The consequences for ethical violations vary widely. In the academic laboratory, lapses in safety ethics may have few immediate repercussions (as opposed to the severe consequences for research fraud, ranging from embarrassing paper retractions to sanctions that cost the violator his or her career), but if actual physical damage to property or (in particular) persons occurs, concrete negative outcomes are more common, including internal sanctions, professional censure, and civil and criminal charges.

An example of the last is the case of Prof. Alain Louati, chemist at the École Normale Superieur de Chemie (National School of Chemistry of Mulhouse; ENSCMu; Mulouse, FR). On 24 March 2006, shortly after noon, an explosion of ethylene gas occurred in ENSCMu's Building 5, a research laboratory building; ironically, the explosion started in a laboratory identified as "Safety and Chemical Engineering."

The explosion resulted from failure of substandard (allegedly recycled and "decayed") rubber tubing connected to a pressure regulator on an ethylene cylinder. The investigator, Dr. Louati, had left for lunch, leaving the experiment to which the tubing was connected off but leaving the cylinder full open. (It is advisable not to open gas cylinders fully unless necessary to achieve the desired flow; only a quarter-turn in this case might have sufficed.) The old tubing was friction-fit to a hose barb on the pressure regulator, and during lunchtime, the connection failed, and ethylene leaked continuously into the lab atmosphere.

After some time, a flammable mixture formed (ethylene is flammable across a range of approximately 3–36 volume % in air) and an unknown ignition source triggered the explosion, quickly releasing an estimated 350 MJ of energy. The overpressure wave from this "confined deflagration" caused damage to the building and campus observable as far as 200 m away; debris was found up to 100 m away from the site. The resulting fire burned for over 5 hours and required about 150 firefighters to suppress [32].

The blast injured 21 people elsewhere in the building (from overpressure caused by the explosion, wall collapses, etc.), including serious wounds to a female high-school student serving as an intern in the department. The building was damaged to the point that complete rebuilding was required. Most seriously, it killed photochemist Prof. Dominique Burget, who was eating his lunch in the office immediately above the laboratory [33].

In 2010, Louati was judged to be the "involuntary cause of death" of Prof. Burget due to "clumsiness, imprudence, inattention," and neglecting prudent safety precautions [34]. He eventually received an 18-month suspended prison sentence for "homicide and unintentional injuries through negligence" [35, 36]. He was also fined nearly €8,000.

3 Assessing and controlling risk

The concept of "risk" means different things to different people, and it even means different things to the same person in different contexts. The concept is fluid and dynamic, and in this chapter we will discuss what risk is in the context of laboratory experimentation, what it is not, and how to estimate it.

The control of risk is the core of laboratory safety, because it is the means through which we measure success. Laboratory safety aims at control of both everyday risks (cutting one's finger on some broken glass) and catastrophic risks (cutting one's finger on glass contaminated with Ebola hemorrhagic fever). Because catastrophic incidents are high-consequence/low-probability events (see below), very few, if any, research institutions have enough of them to develop good statistics. We therefore rely on measures of risk as a proxy for the actual number of incidents that will occur (or persons who will become injured).

3.1 Distinguishing hazard from risk

Risk and hazard are not the same thing. Although we speak of them as synonyms in many cases, mixing the concepts is not correct. A **hazard** is a property that carries with it the possibility of injury, death, or property damage (whether that "property damage" is destruction of physical property, the degradation of the capability of an organization to perform work—like a lab which is shut down by the effects of a fire, or more intangible damage such as besmirching the organization's reputation.) Hazards in themselves are simply mental constructs—ideas (e.g., "dropping this large book might hurt my feet")—and they require some sort of way to become realized ("I broke my toe when I dropped this book on my feet.")

Laymen (and dictionaries) often conflate the concepts of hazard and risk. To professionals, a **risk** is a measure of the "seriousness" of a hazard—a combination of the consequence(s) and probability of realizing the hazard in some fashion. Colloquially, we often speak of "risk" as a proxy for one or the other—saying "nuclear power is risky" when we mean "nuclear power accidents could have very severe consequences," (despite the low frequency of serious accidents), and saying "backing out of a parking space too fast is risky" when we mean "there is a high probability of a minor accident if one backs out of a parking space too quickly" (despite the fact that neither car involved is likely to suffer serious damage).

3.2 Simple methods for estimating risk

Estimating the risk of an incident requires estimating both the potential severity of the incident and the likelihood of it occurring. In laboratory applications, both estimates are best done **qualitatively**. This is because the **quantitative** estimation of risk generally requires expertise and data not available in the research environment.

In industrial applications, there may be many years of operating experience available upon which to base consequence and probability estimates—such-and-such a type of valve is known to fail approximately 1 time every 5 years, for example—but in a research lab, many activities are completely novel—neither consequence nor probability data are available. Even for activities that are routine (e.g., changing the carbon dioxide cylinder on a controlled-atmosphere incubator), it is rare that high-quality data on mishaps and failures is available.

One very commonly used way of "estimating" risk is to simply use the experimenter's instinctive opinion: is the risk of a particular accident scenario "high," "medium," or "low?" A "high" risk would equate to "we should do something about this immediately," before proceeding further, while a "medium" risk would mean that action should be taken to mitigate the risk as resources allow. "Low" risks do not require action. This "gut feeling" is obviously a poor way to estimate risk, as it depends critically on the experience of the person doing the estimation, and it is not calibrated to produce similar estimates between researchers.

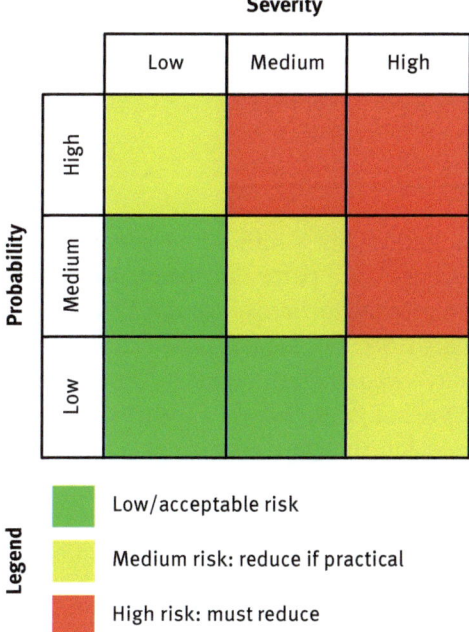

Figure 3.1: 3-level consequence-probability risk matrix.

Qualitative ratings of risk are affected by availability bias. If the researcher has experienced or heard of an incident occurring from a particular material, piece of equipment, or laboratory operation, risk tends to be overestimated. The converse is also true: if an experimenter has never heard of or considered a certain hazard scenario, it will tend to be summarily dismissed as nearly impossible. The high-consequence/low-probability nature of most serious laboratory accidents means that any given experimenter is unlikely to have experienced (or possibly even heard of) such an incident, and so there is a natural tendency to underestimate the risk of certain hazard scenarios.

Another common (and somewhat more defensible) method of rating risk involves assigning numerical ratings to both consequence and probability and constructing a matrix as shown in Figure 3.1. The number of levels used and the exact criteria for assigning ratings varies widely. Because this method is more systematic, it gives the impression of improved accuracy; this impression is often false. In many cases, the consequence and probability ratings are no more defensible than the "high/medium/low" gut feel method.

Consider the 5-level rating system for probability and consequence illustrated in Table 3.1.

Table 3.1: Example five-level probability-consequence rating system

Rating	Consequence severity	Probability of occurrence
1	Minor first-aid injury	1/100 yr
2	Injury requiring medical assistance	1/25 yr
3	Injury requiring hospital admission	1/10 yr
4	Permanent disability/maiming	1/5 yr
5	Death	1/yr

This set of criteria appears initially to be quantitative, but it suffers from serious flaws. First, estimating the "worst possible consequence" of a particular hazard scenario is quite difficult—most laboratory accidents can escalate to "researcher dead," although actual occurrence of a fatality may be farfetched. Second, there is usually insufficient data to produce a reliable estimate of probability—who can say how likely a researcher is to drop a beaker full of acid on the floor? The method does, though, appear to possess a certain degree of quantitative accuracy and precision, though, and so it is popular among attorneys and managers because it is relatively defensible in the event of an accident.

3.3 A semiquantitative method for risk estimation in the laboratory

One way to address the flaws of quantitative and qualitative risk assessments is to construct a system that "bakes in" some level of professional judgement—a system that removes the necessity for a researcher to make estimates he or she has neither data nor experience in making. We can do this by using semiquantitative estimates of severity and probability. The system presented in summary below was created by the author and N. J. Leon [37].

In a semiquantitative system, we estimate the severity and probability of a hazard scenario given the general features of the experiment or activity. This reduces the researcher's job to identifying those general features in a "check the box" sort of manner—in fact, a semiquantitative system can be easily implemented in a tabular or checklist format.

Providing a table or checklist of hazardous conditions simplifies the process of risk assessment, particularly for personnel who are not experienced with the hazard under study. This makes such a system ideal for use with junior researchers, who routinely need to make risk judgments in situations where they lack expertise.

The system presented below is discussed in terms of its theoretical basis; [37] contains detailed instructions and checklists for practical implementation. Severity and probability of undesirable consequences is represented on a 0–5 scale, with 0 indicating no effect and 5 being the worst-possible result. The system is novel in that severity and probability ratings are **summed** to establish a risk rating; more commonly in systems of this type, risk is calculated as a product of severity and probability. An additional novel feature is that explicit credit is given for the nature of risk controls in use—mitigation; this addition allows the user to determine whether existing or proposed risk controls have a significant effect on the risk.

Estimating risk in the system is straightforward. For a given experiment, the most reasonable estimates of severity and probability are made from Tables 3.2 and 3.3. For purposes of this discussion, we assume that the researcher will make the estimate; [37] contains aids to standardize this process and provide the new researcher with assistance in making the estimate. The existing risk control/mitigation measures are also assessed and a numerical rank calculated based on Table 3.4.

Table 3.2: Severity rating system

Ranking	Consequence severity
0	No harmful effect
1	Equipment damaged but still usable (possibly at reduced function)
	Experiment run ruined but repeatable
2	Equipment damaged and unusable without repair/replacement
	Experiment run ruined and not repeatable
3	Personnel injury
	Mild environmental consequence (material escapes lab but substantial public exposure unlikely)
	Property damage limited to laboratory
4	Personnel maimed (non-healable injury with loss of function)
	Moderate environmental consequence (possible injury to public or environment)
	Property damage limited to building
5	Personnel killed
	Major environmental consequence (public response required)
	Large-scale damage (multiple structures)

Risk is then simply estimated as Severity + Probability − Mitigation. Appropriate action can then be taken based on the risk level—see Table 3.5.

Table 3.3: Probability rating system

Ranking	Probability
0	Incident not possible
1	Incident hypothetical/highly unlikely
2	Incident reasonably possible (might happen somewhere during researcher's career)
3	Incident might occur 1/several years (might happen to researcher during his/her career)
4	Incident might occur 1/yr (might happen to researcher occasionally)
5	Incident occurs regularly

Table 3.4: Mitigation rating system

Credit	Mitigation type
0	No mitigation and hazard not "immediately obvious"
1	Hazard "immediately obvious" at all times PPE specified but not well-enforced
2	Warning signs, visual inspection PPE enforced Weak administrative controls (mitigate/detect hazard)
3	Training, SOPs, maintenance Strong administrative controls (prevent hazard)
4	Design features mitigate event (e.g. defeatable interlock) Weak engineering controls (mitigate/detect hazard)
5	Design features eliminate event (e.g, non-defeatable interlock) Strong engineering controls (prevent hazard)

Table 3.5: Risk ranking and actions

Risk ranking	Risk evaluation and action
0–3	Acceptable risk; proceed immediately
4	Questionable; consult principal investigator or supervisor
5	Review required; create Standard Operating Procedure/experimental protocol and proposal for additional risk controls for review by principal investigator or supervisor
6	Marginally-unacceptable; consult safety professional (e.g., Health & Safety Department) for risk reduction measures
7–10	Unacceptable risk; do not proceed without adding mitigation

3.4 Risk assessment exercises

1. Using the scale in Table 3.3, estimate the probability of the following laboratory accident scenarios, assuming you use the relevant materials frequently:
   A. Burn hand on hot plate
   B. Inject self with mouse carcinoma cells
   C. Laboratory microwave leaks radiation
   D. Electric shock from frayed cord
   E. Spill 5 mL ethanol
   F. Spill 4 L ethanol
   G. Accidentally view of laser beam
   H. Spill 0.5 g ethidium bromide during powder weighing
2. Using the scale in Table 3.2, estimate the severity of each of the incidents in Exercise 1.
3. Using the system in Tables 3.2–3.5, estimate the risk of the following accident scenarios:
   A. Chemical synthesis of a new mercury compound, conducted in sealed apparatus in a chemical fume hood (a cabinet for chemical containment) and using an explicit experimental protocol that includes safety measures. Note that mercury compounds are frequently strongly neurotoxic.
   B. Use of a wind tunnel having a 50 hp electric motor and no guard over the drive belts. Warning signs are posted to advise researchers of the hazard from the belts.
   C. X-ray diffraction measurements taken using synchrotron radiation. Synchrotrons produce intense beams of X-rays. The radiation area is shielded with an interlocked shield (i.e., if the shield is opened, the X-rays are immediately shut off) and strong controls are in place to prevent exposure.
   D. Growing 5 gallons of HIV samples for use by other researchers in a bioreactor. The bioreactor system is sealed, but open handling of the samples is necessary during dispensing.

4 Hazard and risk controls

Controlling hazards and the resultant risks is the foundation of laboratory safety. In the laboratory, researchers conduct experiments—and the point of an experiment is to control the experiment and its environment so that only certain variables are changed: the scientific method. Poor risk control equates to poor control of experimental variables and the laboratory environment—meaning poor science.

Good science is safe science.

4.1 The hazard control process

Hazard control is not a disjointed set of safety rules and policies; it is an organized process of recognizing and controlling risks. Organizations, including laboratories, that approach safety from a rules/policies standpoint are operating in a reactive mode, creating protection layers with large "holes," in the language of Reason's model of accidents. (See section 1.2.1, above.) It is far preferable to use a proactive method of seeking out and controlling hazards and their associated risks.

There are four main steps to the hazard control process:
1. Hazard identification
2. Risk screening
3. Hazard analysis
4. Hazard control

4.1.1 Hazard identification

The first step in controlling a hazard is to recognize its presence. Note, though, that a hazard is an abstract thing: turpentine burns easily, for example. This does not mean that a fire occurs every time turpentine is used.

We will distinguish the term **hazard**, a dangerous property of a material or procedure, from a **hazard scenario**, the chain of events through which a hazard is realized. This chain of events (or more realistically, a network of causes and conditions) is the mechanism through which the hazard causes injury or damage. Using turpentine while smoking will likely lead to a fire or small explosion—this is a reasonable hazard scenario (and one that would require further analysis).

Hazard identification techniques vary from simple and **ad hoc** to complex and highly-organized. (We will learn more about hazard identification in chapter 5 below.) An example of a simple hazard identification technique would be using a checklist of common hazards similar to that in chapter 16 below; more complex hazard identification is generally a component of an organized hazard analysis technique, such as the **Hazard and Operability Study (HAZOP)** [38].

4.1.2 Risk screening

Once the presence of a hazard is known, the second step is to estimate the risk posed by the hazard. Most experiments have a plethora of associated hazards, each of which is capable (under sometimes highly-contrived circumstances) of escalating to "experimenter killed." It is **essential** for an experimenter to have the ability to quickly and accurately determine which "researcher-dies-a-horrible-death" scenarios pose significant risk and which do not. Resources available for safety—time, money, and other items—are limited; no researcher wishes to spend a large amount of money on safety equipment to avoid an accident whose risk is actually low, particularly if there may be much-higher risks present.

Consider a common experimental operation in a mechanical engineering laboratory: machining a part with a lathe. Several hazards come to mind immediately: The researcher could:

1. Catch his or her hair on a wrap point (see section 6.1.5 below) and be pulled into the machine, dying of suffocation because there was no one else in the shop to offer assistance;
2. Be electrocuted by a short circuit in the motor;
3. Be struck by a soccer ball which crashes through a nearby window and die of a subdural hematoma.

Each of these scenarios, while hazardous, even deadly, is **not** of equal probability, and thus is not of equal **risk**.

The first risk is quite high: it has occurred many times in machine shops, including in 2011 at Yale University, killing a student (See section 1.1.3 above). The second risk is moderate: motors do experience short-circuits, particularly if they are poorly maintained, but standard grounding practices act to protect the user in most cases. The third risk is possible, but highly improbable: it is a low risk. Unless the machine shop is located immediately next to an athletic field and stray balls regularly fly into the room, taking some kind of action to mitigate the risk is likely to be a waste of time and money, distracting attention from the more serious first risk.

It is necessary to understand exactly how one might be exposed to a hazard in order to accurately estimate the risk. Without spending the time and energy to do a full hazard analysis (see section 4.1.3 below), a useful technique is quickly to identify a "worst credible case" and estimate the risk from that, using it as a proxy for the aggregate risk from all possible scenarios involving that hazard. The worst credible case is the most severe risk that, in the researcher's opinion, is **reasonably capable of occurring**.

In the case of the lathe, one can envision several possible scenarios involving physical hazards: eye injury from flying chips; loss of a finger or hand to an inexpertly-handled tool; or the hair-caught-in-wrap-point scenario mentioned above. The last is obviously the "worst credible case."

4.1.3 Hazard analysis

In the hazard analysis process, we consider abstract hazards (e.g., flying chips, in our lathe example) and attempt to determine how these hazards may be realized. There are a wide variety of techniques used for hazard analysis, but all of them involve somehow connecting the hazard to the "real world" by generating hazard scenarios (e.g., flying chips striking the operator in the eyes). The various hazard analysis techniques all answer the same question—"What are the consequences of a particular hazard scenario, and how can one protect against them?"—but they have different applications, depending on how the hazard scenarios are generated and how deeply they are analyzed:
- The **"What If?"** method creates scenarios through brainstorming, often by experts;
- The **Checklist** method verifies that a certain set of controls are in place or assists in identification of hazards present;
- The **Hazard and Operability** Study (HAZOP) uses a set of guide words that are applied to the hazard to guide scenario generation;
- **Job Hazard Analysis** (JHA)/**Job Safety Analysis** (JSA) is a simplified method often useful on lower-risk procedures that can be broken up into individual steps;
- **Failure Modes and Effects Analysis** (FMEA) specifically studies the failures of a particular component (not necessarily a physical one).

Most, if not all, hazard analysis techniques include estimation of the risk of the scenario, although this is sometimes implicit. Many methods, such as HAZOP or "What-If?," generate a large number of possible scenarios, not all of which will be risky enough to require mitigation (or even be realistic). Including specific risk estimation of some kind allows the analyst to concentrate attention on the most risky cases; a key concept of modern hazard control is **"worst-first,"** meaning, "concentrate resources on the most risky scenarios."

4.1.4 Hazard control

Only after the important risks in the experiment, process, or procedure are clear should one start to consider appropriate controls. While a suitable control may in many cases be "obvious," it is essential to give full consideration to possible controls that might reduce the risk from a particular scenario. It is often the case that on second consideration, a "non-obvious" hazard control solution becomes clear—and it is often superior to the reflexive first idea. For theoretical and practical reasons, also, certain hazard control solutions are considered better than others. (See section 4.2.2 below.)

In the case of the lathe used above as an example, there are several controls that might prevent eye damage from flying chips:
- The operator can wear safety glasses and face shield;
- A clear shield could be installed between the operator and the workpiece;
- The lathe could be operated remotely, with sufficient distance between operator and tool that chip impact is unlikely;
- Specific training and procedures could be required before an operator is permitted to use the lathe.

4.2 Classifying hazard controls

Hazard controls may be classified using two different schemes, concentrating on what they do (mode of performance), and what they are (inherent nature). Consider the three ideas from the lathe example above.

4.2.1 Functional classification of hazard controls

If we consider the **function** of the hazard control, we can classify them in three categories:
- Those which **prevent** the hazard scenario from occurring;
- Those which **mitigate** the consequences of the scenario; and
- Those which **detect** that the hazard or one of its consequences are present.

The order given above—prevention, mitigation, and detection—is usually the preferred order. Note, though, that we evaluate hazard controls according to several criteria:
1. **Is the control effective?** Does it actually reduce risk from the scenario? Safety glasses, certainly, reduce the probability that the operator's eyes will be struck by flying chips, although they do not reduce it to zero.
2. **How much does it reduce risk?** Some measures may offer only a little reduction in risk. A face shield alone, for example, does not provide much protection to the eye. Face shields should be used only as secondary protection for safety eyewear—they protect the face, not the eyes.
3. **Is it defeatable?** If a careless or pressured operator can disable or omit the use of the control, net risk reduction is smaller. Safety glasses, unfortunately, fall into the "defeatable control" category.
4. **Does it introduce any new hazards?** Sometimes hazard controls, especially if they are poorly executed, can **create** hazards of their own. An inexpensive pair of safety glasses may be of such low optical quality that the operator is more likely to make a mistake, for example.
5. **Is the cost low?** Only a certain amount of resources, including money, is available for ensuring safety, and making efficient use of funds and personnel time is

essential. Remote operation of the lathe, the last possible control above, would likely be very effective in reducing risk, but at a very high cost.

In general, though, the order of preference is **prevention—mitigation—detection.** Prevention removes the hazard entirely. Mitigation simply reduces one or more of what may be several consequences—the hazard scenario has already begun. Detection is useless without appropriate plans and training on what to do in the event detectors indicate a dangerous condition.

4.2.2 Traditional hierarchy of controls

While the function of the possible hazard control is, in the author's opinion, the most important factor in choosing hazard controls, industrial hygienists and safety professionals also classify controls by **their nature**, that is, what they are (as opposed to what they do):
- **Engineering controls** are physical barriers or controls, such as the lathe shield mentioned above, that prevent an incident or interrupt the chain of resulting consequences;
- **Administrative controls** are procedural restrictions, work rules, training, etc. that reduce the incidence and/or severity of an incident;
- **Personal protective equipment (PPE)** is physical equipment worn on the body to reduce (one hopes) the severity of an incident.

There is a definite order of preference to this **hierarchy of controls**: engineering controls are considered superior to administrative controls, which are considered superior to PPE. In theory, this is mainly for reasons of reliability. Properly maintained, an engineering control is more likely to be effective than an administrative control, which relies directly upon people to follow it. PPE is dependent upon proper use by the protected individual (which is notoriously difficult to enforce) and tends to be less than completely reliable. (For example, a strong splash of liquid can loosen or even knock off a pair of chemical splash goggles, obviating their usefulness.)

One must use caution in applying the hierarchy of controls. Blindly choosing an engineering control over an administrative one does **not** guarantee that the control will be more effective because it does not depend on human intervention. Engineering controls depend critically on human beings at several points:
- **They must be properly designed.** Guards, for example, must be placed further and further from the hazard as any gaps provided (e.g., to pass workpieces through the guard) increase.
- **They must be properly constructed.** Shoddy or improper installation of an engineering control can be **worse** than no engineering control at all. (The author once encountered a "chemical fume hood" that was venting its exhaust into the lab.)

- **They must be properly maintained.** A loose guard can become nearly useless.
- **They must be properly used.** Self-closing valves, used to limit possible leaks from equipment, for example, are often wired open by users to avoid the inconvenience of holding them open when necessary.

4.2.3 Creativity in hazard control

Hazard control is a science, and like all sciences, it depends critically on the scientist (or engineer's) ability to switch back and forth between two modes of thought: the creative and the analytical.

Creative thought modes permit the safety analyst to identify the broadest range of hazards and possible solutions. The "What-If?" method of analysis (see chapter 12 below), for example, relies upon brainstorming techniques to generate questions to be analyzed. Over-reliance on creative thought, on the other hand, can lead to inadequate analysis—hazard scenarios that are not thought through completely, or hazard controls that, while clever, are not as effective as possible.

Analytical thinking, on the other hand, is essential for a full inspection of each possible hazard scenario. Accidents happen as a chain or net of events, one leading to another, and creative thinking techniques such as brainstorming are not well-adapted to analysis. The linear analytical mode of thought, if applied universally to safety analysis, leads to "tunnel vision," the inability to see alternative causes, consequences, and solutions.

The author has observed that the ability to "dream up" creative safety questions and imaginative solutions **declines with education and experience**. This may be because as we learn, whether by schoolwork or experience, we increasingly learn that phenomena progress in certain predictable ways: turning a piece of wood on a lathe progresses by gradual removal of material as the tool is gently applied to the workpiece. This familiarity with the "typical" chain of events seems to discourage the brain from considering alternative possible courses; in the case of the lathe, accidentally applying the tool too firmly to the workpiece can lead to kickback—and the operator being struck in the face by the sharp tool. The author recommends that readers obtain formal creativity training or review books such as **Mindmapping** [39] or **A Whack on the Side of the Head** [40] to combat this effect.

4.3 Exercises: Hazard control

1. Into what categories may hazard controls be classified **by their function?** (Choose one.)
   a. Engineering controls, administrative controls, personal protective equipment

b. Detection, administration, and prevention
c. Prevention, mitigation, and detection
d. Engineering, administration, and protection
2. Which of the following are elements of the hazard control process? (Check all that apply.)
a. Identifying the hazard
b. Controlling the hazard
c. Designing engineering controls
d. Safety budget management
3. Consider the following scenarios. In each, identify one engineering control and one administrative control that might apply to the situation.
a. A researcher is using ethidium bromide (a DNA stain) to conduct DNA fingerprinting experiments. Being splashed with this chemical is dangerous.
b. A researcher is operating an experiment using a high-voltage power supply. The power supply is open and energized, which could lead to a deadly shock.
c. A researcher is aligning a high-power laser on an optical table. The researcher and bystanders have open access to the beam, which could lead to serious eye injury.
4. You are attempting to protect against a fire in a lab. Match the proposed control with its function and determine what type (engineering, administrative, etc.) of control it is.

Controls
1. Installing a fire sprinkler system to extinguish fires.
2. Removing all combustible items from the lab.
3. Using a heat sensor that sounds an alarm when temperatures reach 120°C.
4. Providing fire-retardant lab coats to the researchers.

Functions
A. Detection
B. Prevention
C. Mitigation

Type/nature
a. Engineering control
b. Administrative control
c. Personal protective equipment

Part II: Hazard classes and control methods

5 Hazard identification methods

The first step in controlling a laboratory safety risk is to identify the hazard. Methods and sources to accomplish this vary widely—some hazards may be identified from personal experience without recourse to other resources, while others may require consulting simple checklists or performing literature (and even laboratory) research. In this chapter, we will briefly discuss methods for identifying laboratory safety hazards, along with their strengths and weaknesses.

5.1 Brainstorming, mind-mapping, and other creative methods

Many hazards are difficult to see at first. While an electric cord laid across a walkway is a fairly obvious trip hazard, there is also a fire hazard due to possible damage to the cord from being walked upon. A hazard is not low-risk simply because it is non-obvious or obscure. Creative problem-solving and idea-generation methods are often useful in "stepping back" to evaluate a system from different perspectives. Two of such methods are brainstorming and mind-mapping.

Brainstorming is a common technique used to generate possible issues and solutions in a variety of systems, ranging from laboratory experiments to business decisions. It may be performed singly or in groups, depending on the depth of analysis required, the breadth of expertise needed, and the resources available. In safety evaluations, brainstorming is most useful when performed as a team—team members, particularly if they have different backgrounds, areas of expertise, and experience levels, tend to generate a broader range of ideas than a single person.

In a brainstorming session, team members attempt to generate as many ideas as possible. There should be no censorship of ideas as "stupid," "unworkable," or the like. Oftentimes, a "stupid" or "silly" idea from one analyst will spark a very useful and reasonable idea in another. In a second phase, the group identifies the ideas that are actually relevant and useful.

Brainstorming requires tight collaboration among participants, and personality differences or cross-cultural issues may interfere with this collaboration, leading to "groupthink" phenomena where aggressive or ranking members dominate the discussion. See [40] for more details on brainstorming.

Mind-mapping is a method for free-form idea generation and analysis. Starting with a central concept, which may be broad in scope ("spilled oil at sea") or narrow ("electrical failures in power supply"), the investigator uses free-association, brainstorming, and other techniques to generate "connected" ideas, representing them graphically as a 2-dimensional "map". Mind-mapping may be used by individual researchers or by groups. In safety, it is primarily useful as a method for generating **and organizing** possible issues, outcomes, and effects of experimental procedures in a manner that is free-form but still organized. See [39] for further detail.

5.2 Checklists

When a laboratory situation is common and predictable—setting up a compressed gas cylinder for use, for example—many, if not most, of the hazards are predictable as well. In such cases, a **checklist** is an appropriate method for identifying hazards and hazard scenarios. The checklist generally contains the major known hazards of the situation or procedure, and the user is merely required to "check the boxes," or make a brief evaluation of whether that hazard is appropriately controlled. Sometimes, the checklist may list hazard controls instead of hazards; this is common when the hazards of the situation are well-known.

Checklists are also useful in conducting high-level screenings of research and research programs, to identify general areas that require additional scrutiny. Chapter 16 contains such a checklist, used at the author's institution for rapid risk assessment of experimental protocols and as a starting point for reviews of new research programs—particularly those of new faculty.

Finally, checklists can be crucial as a method of documenting experimental protocols and procedures. The most famous checklist uses of this type are in aviation, where a pilot is typically equipped with checklists for each operation, normal and emergency, containing steps to ensure that crucial actions are not omitted or performed out of order. More recently, the checklist method has been applied to medicine, reducing hospital errors significantly [41].

The main weakness of checklists for hazard identification is the tendency on the part of the user to assume that because a hazard is not on the checklist, it is not present in the system—essentially, the user becomes a "checklist zombie," mechanically checking boxes and not actively evaluating the experimental system for other hazards. This can lead to a perfect score on a checklist for an experiment with blatantly obvious hazards specific to the installation or use of the apparatus unrecognized and thus uncontrolled.

5.3 Reference books

It is unlikely that most researchers will consult the safety literature during an evaluation of a routine laboratory procedure. Nevertheless, careful literature review can uncover hazards recognized a long time ago and forgotten since—this effect is partly responsible for the death of Dr. Karen Wetterhahn from dimethylmercury poisoning in early 1997. (See section 1.1.2 above.)

A short, annotated list of secondary safety literature resources appears in Chapter 19. It contains some of the principal "entry points" into safety literature.

In particular, literature methods-and-materials sections rarely list safety considerations. The experimenters may have used the experimental apparatus described or conducted a synthesis step 13 times until they successfully completed it without an explosion, but such facts may never appear in the paper. There is a movement among chemical journals (such as the **Journal of Chemical Education**) to require safety information to be disclosed alongside other methodological data, but this will never help older literature. The author has personally encountered literature formulations (for electropolishing solutions) capable of self-combustion or possibly detonation. (See section 13.1 below.) Always verify the safety of machinery, laser models and setups, and chemical mixtures among other things before contemplating replication of a literature experiment. If the original author is living, consulting him or her regarding safety issues encountered is strongly suggested.

Note that Internet reference sites such as ChemSpider (http://www.chemspider.org) and Wikipedia (http://www.wikipedia.org) are not reliable sources of safety information—discussions with staff of each website indicate that each has made an explicit editorial decision not to include curated safety information. Any "lower flammable limit" or "LD_{50} toxicity levels" (see below for definitions of these terms) found in such sources should be cross-checked against the literature—and a full literature search performed, not just a cross-check against any footnote provided in the secondary source.

5.4 Regulations and standards

While most people think of laws, regulations, and standards as "safety rules" that place restrictions on what work can be done in the lab, they can also be useful sources of proactive safety information. To clarify the use of terms:
- A **law** is a restriction or obligation imposed by the local or national legislative authorities (e.g., the US Congress). Laws relating to occupational safety are generally vague and delegate much of their implementation to executive authorities.
- A **regulation** is a restriction or obligation, typically more detailed than a law but authorized by it, imposed by the local civil authorities (e.g., the US Department of Labor). Regulations generally have the force of law; violating them can lead to fines and imprisonment.
- A **standard** is a voluntary statement of good practices recognized by a particular technical community, for example, users of cooling equipment or lasers. Some standards specify safety precautions (for example, see [42] or [43]). While voluntary, many nations can hold researchers or their institutions liable for civil or criminal violations through general duty-of-care regulations. (The US Occupational Safety and Health Administration—OSHA—refers to its technical regulations as Standards, confusingly. Compliance with OSHA Standards is mandatory for US employers.)

Laws, regulations, and standards are typically **prescriptive** in nature: they tell the user what to do without necessarily explaining why. Some expertise is often required to determine the reason for a requirement and whether it actually applies to the situation—if the hazard that the regulatory requirement is intended to address is not present, an argument can be made to authorities that it is not necessary to comply with the requirement. Expert advice (or sometimes consultation with the regulatory body in questions) may be necessary, and sometimes handbooks are available with interpretive material to aid in application of a complex standard.

Another reason technical expertise is sometimes necessary to interpret regulatory and standards requirements is that they can conflict with one another, creating situations where it is possible to meet only one of two applicable requirements.

5.5 Real-life hazard identification

Ideally, an experiment has two or more possible outcomes, and the investigator's purpose is to objectively evaluate the results to support or disprove his or her hypothesis. In the real world, the researcher has a vested interest: he or she is **expecting the experiment to progress in a particular manner.** There is a cognitive bias toward giving more weight to evidence that confirms the researcher's expectations and discounting (or even ignoring) contrary evidence. This "seeing only what you expect to see" can be detrimental to the quality of research (occasionally leading to cases of unethical behavior such as data falsification) and can also be extremely dangerous. In fixating on the expected behavior of the experimental system, it becomes difficult or impossible to even **think** about unexpected behavior. It is essential for researchers, both for the quality of their science and for the safety of themselves and others, to evaluate in advance all foreseeable consequences of an experiment.

The author once experienced a fire in the lab involving triethylborane, a compound that spontaneously ignites in air. The fire started as a result of a misjudgment—the author's trust in the chemist who synthesized the raw chemical, who claimed to have treated the mixture with suitable reagents to remove the triethylborane (a by-product of the reaction). In fact, he had "quenched" the reaction inappropriately, leaving a substantial amount of triethylborane in solution. When the author's laboratory assistant distilled the reaction mixture to isolate the product, the triethylborane became concentrated in the still bottoms (that part of the mixture left in the distillation apparatus when the product of interest has been removed) and ignited. This fire was foreseeable, and the result of an **assumption** that a portion of the experiment (a step carried out by a graduate student in another research group, in this case) would occur exactly as specified.

The author also learned through this experience that most people do not react well in emergencies. The undergraduate lab assistant pulled the flaming flask out of the fume hood, marched it across the lab, and presented it, exclaiming, "Dr. Kuespert! It's on fire!"

5.6 Exercises: Hazard identification

1. Using the checklist provided in chapter 16, evaluate a laboratory to which you have access and determine which general hazards are present.
2. Select up to four hazards obtained from the laboratory evaluation in Exercise 1, above, and use creative methods such as brainstorming and mind mapping to identify several possible hazard exposure scenarios for each.

6 Physical hazards

As mentioned in section 1.3.1 above, "physical hazards" is a broad default category, so it is impossible to sum up all physical hazards with a simple, straightforward definition. Most broadly, though, a physical hazard involves contact with some form of physical energy—being struck by a bolt thrown off a machine, being struck by the pressure wave of a chemical explosion, or being burned by heat transferred through radiation while standing near a piece of ceramic still white-hot from the kiln. In this chapter, we will cover a selection of the most common and most risky physical hazards found in the research laboratory. The list of all possible physical hazards, though, is nearly endless.

6.1 Mechanical hazards

Mechanical hazards come from moving parts, whether on a shop machine such as a lathe or from shrapnel thrown out by an over-pressurized air tank. Most mechanical hazards are easy to identify, since (apart from the flying-shrapnel example) they appear at discrete points on an apparatus and can be recognized by an educated researcher. Mechanical hazards also arise from mechanical devices that store energy, such as springs; sudden release of the energy (and the consequent moving parts) usually create the actual hazard.

Evaluate all mechanical devices and tools to identify possible mechanical hazards **before** using.

6.1.1 Pinch points

A **pinch point** is a location where two mechanical members converge in-line with one another. (See Figure 6.1.) If a body part (finger, leg, head) happens to be between the members when they close, it may be crushed; the mechanical advantage inherent in many pinch points means that sometimes only a very small amount of mechanical force must be applied. Only one of the members need actually be moving or be able to move, as in the case of the press shown in Figure 6.2.

6.1.2 Guards and interlocks for mechanical hazards

The principal means of controlling mechanical hazards is to prevent vulnerable body parts from entering the area of the pinch point, wrap point (see section 6.1.5 below), etc., when the mechanism is energized. While this can be done with administrative controls, modern safety practice (and many laws) requires the use of **guards** and **interlocks**.

Figure 6.1: Pinch point schematic.

A guard is a physical barrier that prevents contact with or access to the hazardous area. In a laboratory machine shop, for example, guards enclose the bulk of the grinding wheel in Figure 6.3. While the guard need not completely surround the hazard zone (and in many cases it cannot do so without compromising the equipment's function), it is usually installed far enough away that a person cannot place parts of the body where they will be injured or be drawn into the machine. A 1 cm gap is large enough to allow a finger to enter, but a 15 cm gap is sufficient to reach one's entire arm into the machine. As the gap widens, the guard is placed further away from the actual point of danger [44].

An interlock is a mechanism preventing the equipment from being energized until it, its operator, and any bystanders are in a safe state. Usually, interlocks are electrical in nature—a shaft guard around a large motor, for example, may be connected to a contact switch that prevents starting the motor unless the guard is closed and locked—but they may be mechanical as well. (See Figure 6.4.)

Another common interlocking mechanism is the use of switches, levers, etc., which the equipment's operator must activate in order to start the machine. Some large presses use **two-hand control**, in which the operator must depress and hold two well-separated buttons in order to operate the press; since both hands are busy holding down the buttons, it is impossible to reach into the pinch point while the press is energized. (See Figure 6.5.) Naturally, the two-hand control is placed far enough away from the hazard zone that the operator cannot place his or her head in danger. A similar **two-hand trip** does not require the buttons to be held, but requires additional equipment to make sure the operator's body parts cannot enter the danger zone. A more sophisticated interlock might be installed with a powerful robot arm, perhaps using a proximity sensor or light curtain to de-energize the robot if someone enters its range of motion.

Figure 6.2: Laboratory press with pinch point identified.

6.1.3 Shear points

A **shear point** is similar to a pinch point, but the two members come together out-of-line and with little clearance. (See Figure 6.6.) Shear points are often used deliberately in cutting machines—consider the action of an office paper cutter or pair of scissors. The converging members are often long enough to act as levers, again creating a large mechanical advantage for the equipment and increasing the severity of the hazard. A robot designed with offset joints is a common shear point in labs. (See Figure 6.7.) In many cases, it is very difficult to install guards around shear points, and so this hazard is more commonly controlled with interlocks and administrative controls (such as excluded areas around robot arms).

Figure 6.3: Machinist Richard Middlestadt sharpens a tool on a grinding wheel.

6.1.4 Run-in points

A **run-in point, or in-running nip point,** is a location where a rotating part operates in contact with or in the close vicinity of another part (moving or otherwise). (See Figure 6.8.) Examples would be two meshed gears in a gearbox, as in Figure 6.8, or a rack-and-pinion assembly.

Run-in points need not involve parts in contact. Another type of run-in point, a **squeeze point** (Figure 6.9), is formed when two rollers are in close proximity, such as when flat stock is fed into a machine. An old-fashioned laundry mangle is a classic example of a squeeze point. (See Figure 6.10.)

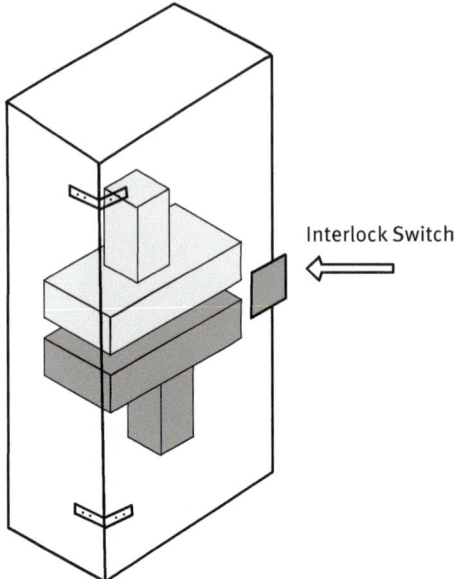

Figure 6.4: Interlocked guard schematic.

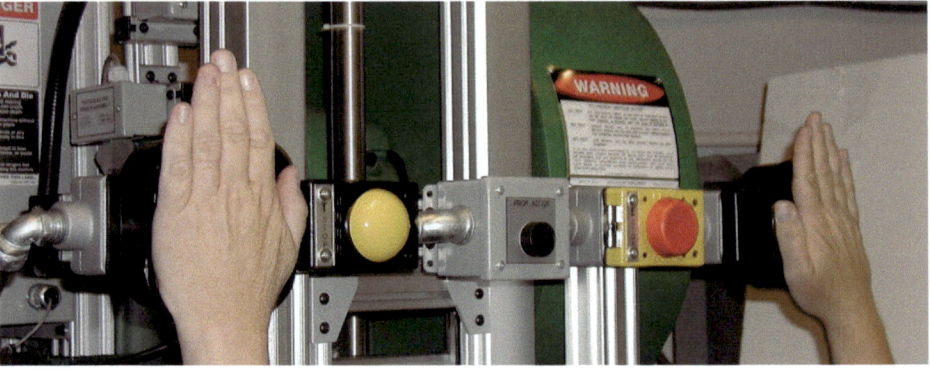

Figure 6.5: Press operator using two-hand control. Photo courtesy Rockford Systems, LLC.

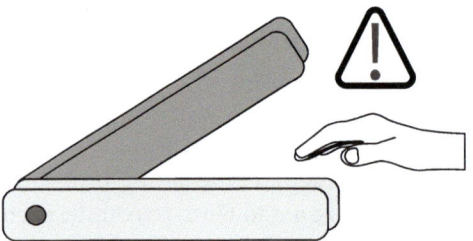

Figure 6.6: Shear point schematic.

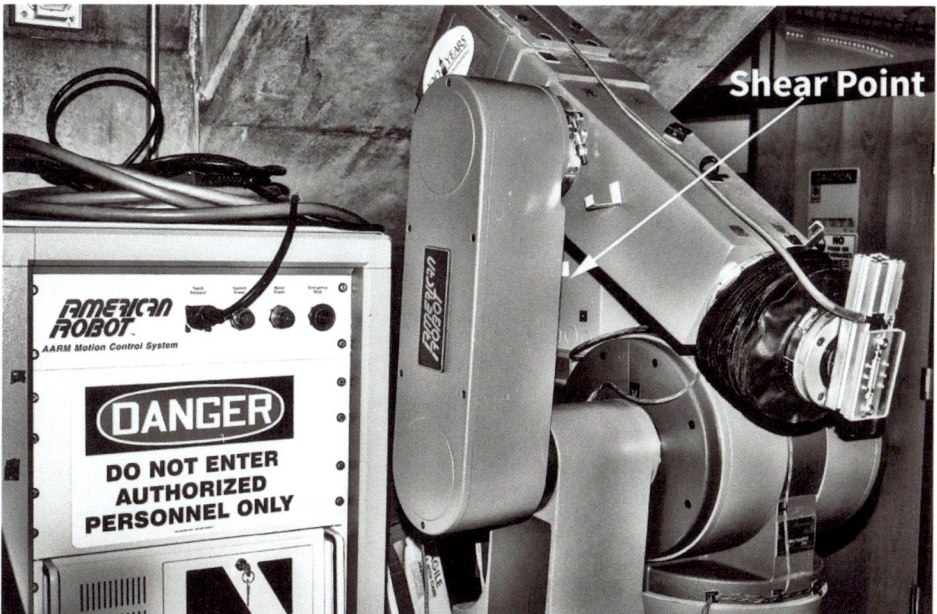

Figure 6.7: Laboratory robot arm with shear point noted.

Figure 6.8: Run-in point schematic (two meshed gears).

6.1.5 Wrap points

Wrap points are points where loose clothing (or even fingers/limbs, etc.) may become wrapped around a rotating part and draw the body into the machinery. A common example in machine shops and materials labs is the lathe, but motor shafts are also common wrap points. In the case of the motors, the wrap point is easily surrounded by a guard, but a lathe is inoperable if completely guarded. The death of Michele Dufault (see section 1.1.3 above) is an example of a typical lathe accident.

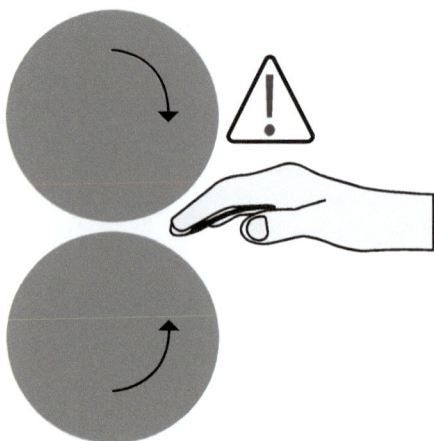

Figure 6.9: Squeeze point schematic (two rollers in contact).

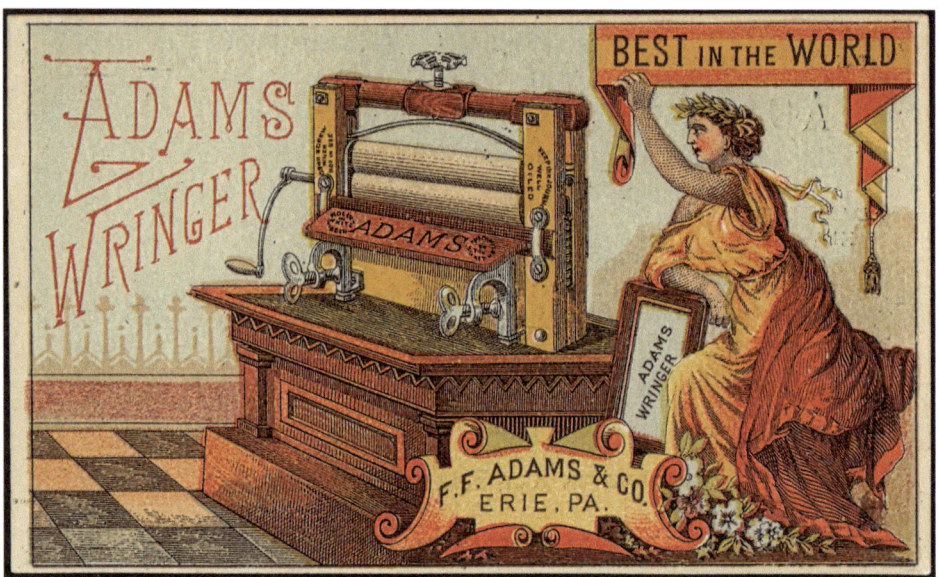

Figure 6.10: Laundry mangle. Photo citation: "Empire Wringer Co., Auburn, N. Y.." Card. 1870. Digital Commonwealth, http://ark.digitalcommonwealth.org/ark:/50959/7m01c310c (accessed January 11, 2016).

Appropriate attire for lab work
Wearing appropriate clothing can be critical in laboratories. Follow these guidelines **before working in any laboratory**:
– **Wear long pants** in preference to shorts, skirts, kilts, etc. For most laboratories, the more the body is covered, the more it is protected from contact with hazardous chemicals, incidental

heat sources, etc. If you must wear shorter clothing, ensure that your knees are covered—this provides some protection to your thighs while you are sitting at the bench. Pants should not have rolled cuffs, since spilled liquid may become trapped.
- **Wear close-fitting clothing.** Loose sleeves, kaftans, etc., may become caught in equipment, knock over chemical bottles, etc. Depending on the hazards of the lab, sleeve length may vary—in a research machine shop, for example, short sleeves are safer, while in a chemistry laboratory, maximum body coverage is key.
- **Avoid synthetic fabrics.** If you are exposed to fire, synthetics tend to ignite more easily and burn faster. They often melt and can provide a great deal of additional fuel—located right up next to your body. Note that fire-retardant synthetics, such as Nomex IIIA® (DuPont) are an exception. Sheri Sangji's acrylic sweater was a contributing factor in her death. (See section 1.1.6 above.)
- **Tie back long hair.** Just like loose sleeves, loose hair may accidentally catch on fire, be dipped in chemicals, knock over containers, or become caught in a machine (such as occurred in the Yale incident mentioned earlier —see section 1.1.3, section 6.1.5 above).
- **Remove jewelry.** This is particularly important in labs containing machines or lasers. In the former case, rings and necklaces may become caught in the equipment, and you could easily lose a finger (or head). In the latter, stray beam reflections off of shiny jewelry can be extremely hazardous and unpredictable. Reagents in use or in the lab atmosphere may corrode jewelry as well.
- **Wear closed-toe shoes.** Such shoes provide some protection against impact—special-purpose safety shoes (often called "steel toes" after their reinforced toes and metatarsal guards) can provide a great deal more. In laboratories that handle liquids, such as chemical reagents or biological cultures, prefer shoes with continuous uppers over porous materials such as canvas tennis shoes.

6.1.6 Clobbering

"Clobbering" is a slang term for being struck by a moving part (it implies being hit quite hard). Standing within the range of motion of a robot arm while it is energized creates a risk of being clobbered. Overhead cranes in engineering laboratories often carry this hazard, as the dangling hook and chain can swing unpredictably during operation.

Eye protection for physical hazards
Safety eye protection comes in many varieties, each matched to a particular hazard. In the case of physical hazards, the most common risk to the eyes is being struck, whether by flying glass from a dropped beaker or by a robot arm that malfunctions. It is **essential** to choose the correct eye protection for all hazards to which you will be exposed. There are two basic types of eye protection appropriate for impact hazards:
- **Safety glasses** are worn just like regular prescription eyeglasses, but they are reinforced to withstand impacts without shattering. Safety glasses are available in a variety of styles, sometimes darkened for outdoor use and sometimes tinted to enhance contrast. Use only safety glasses equipped with side shields or equivalent protection from a wraparound design.

Prescription safety glasses use shatter-resistant lens materials, usually polycarbonate or Trivex®(PPG) and specially-tested frame designs, but they are normally no more costly than traditional frames and lenses. For laboratories short of funds for prescription safety glasses, certain safety glasses are designed to fit **over** the user's regular glasses. Inexpensive "visitor specs" (spectacles) for short-term disposable use are usually constructed this way as well.

- **Impact safety goggles** provide all-around protection since they fit close to the face. Goggles are more suitable than glasses for situations involving smaller pieces of material—grinding, cutting hard materials such as concrete, working with a chisel, etc. Since goggles fog much more easily than glasses, they are typically ventilated; in the case of goggles designed for impact hazards, this normally takes the form of holes drilled in the body of the goggles to permit air circulation. Goggles are available to fit over prescription glasses, and prescription lens inserts are also available for some models.

Impact glasses and goggles are not suitable for other hazards! They offer little protection against chemicals and hot or cold liquids in particular—droplets may penetrate the direct ventilation holes of impact goggles, while liquid splashes may make their way around glasses or strike the forehead and run down into the eyes. **Never** use impact safety glasses or impact goggles to protect against chemical hazards, lasers, etc.

Most safety eyewear in the United States is stamped "Z87" or "Z87+," indicating conformance with an American National Standard (ANSI standard) covering safety eyewear. (The "+" designates stronger glasses suitable for military and other high-impact use.) The equivalent European mark is the well-known "CE" mark.

6.2 Sharps

Sharps in the lab pose an obvious cut and/or puncture hazard. Sharps are involved in many, many laboratory incidents; representative scenarios include:
- A researcher was feeding a glass tube through a rubber stopper in chemistry lab but neglected to lubricate it with grease or glycerin. The tube broke and slipped, completely penetrating the researcher's palm.
- Another researcher attempted to pick up a razor blade that another lab worker had left on the bench top with thumb and forefinger. The researcher received a painful cut under the thumbnail.
- A lab worker operating an old microtome lost concentration and removed a thin slice from her knuckle.
- A biological researcher accidentally injected himself with mouse cancer cells while attempting to inject a live mouse.

The first incident results from an operational error (failure to lubricate the tube), the second from careless laboratory management (not storing a naked blade safely), the third from the inherent properties of equipment (the old microtome's lack of a guard), and the last from a lack of appropriate needles, protective gloves, and technique (shielded needles and puncture-resistant gloves should have been used).

6.2.1 What is a sharp?

The term **sharp** encompasses much more than just syringe needles used in biological work. In general, a sharp is any item capable of cutting or penetrating the skin of a researcher. Many things are laboratory sharps:
- A razor blade or X-Acto®(Elmer's Products, Inc.)-type knife used to cut samples from bulk material in a materials science laboratory;
- A scalpel used for surgery on animal subjects in a zoology laboratory;
- Hypodermic syringes and needles used for transferring liquids in chemistry laboratory;
- Glass Pasteur pipets used to transfer culture medium in a biology lab;
- Broken glassware in nearly any lab;
- Metals with un-filed sharp edges (e.g., from cutting with aviation snips) in a mechanical engineering lab;
- Nails in a civil engineering lab;
- Small electronic components such as resistors and DIP (Dual Inline Package) chips in an electrical engineering lab;

and so forth. Some items are inherently sharp, while others become sharp only after damage.

Sharps are more than just a physical hazard. Once the skin is penetrated, a pathway is created deep into body tissue or even to the bloodstream. Any material on the skin, on the sharp, or on any gloves that were breached now has direct access to the interior of the body, bypassing the skin's role as a protective barrier. Diseases such as tetanus are easily contracted, and toxic materials can be quickly dispersed to organs particularly susceptible to them (such as kidneys and liver).

6.2.2 Sharps handling

Appropriate technique is the most essential factor in avoiding sharps injuries; follow the advice below to minimize your chances of a cut or puncture.

Avoid using sharps whenever possible. For example, instead of using a scalpel to cut slices of tissue, one could use a microtome—modern microtomes are equipped with guards or other safe design features. Choose plasticware over glass where practical to eliminate or greatly reduce the chance of shattering.

Sheathe sharp tools when not in use, where this operation does not create the risk of a cut or puncture. Disposable blades and needles, though, should not normally be re-sheathed or re-capped, since the hand and fingers are typically very close to the sharp while doing so. Probably the most common sharps injury in healthcare occupations (and in biology and chemistry labs) is a puncture while attempting to re-cap a needle. (This can be a career-ending, or even fatal, incident if the needle is contaminated with a pathogen such as hepatitis C.) Self-recapping needles are available for many applications, although more expensive than plain disposables (see Figure 6.11); if this is not an option, plastic hand guards are available to prevent stabbings.

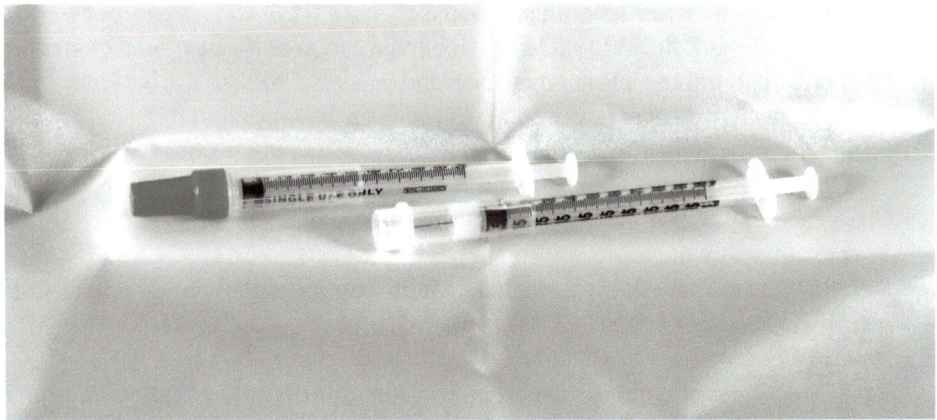

Figure 6.11: Self-sheathing needle. Photo courtesy Sol-Millennium Medical, Inc.

Lubricate glassware with glycerin or water before attempting to insert it into stoppers or gaskets, and use caution during this operation. A classic "beginner's chem lab accident" is a cut or puncture from attempting to force glass tubing through a rubber stopper; oftentimes, the victim ends up with broken glass sticking through his or her palm.

Have a designated area to **store sharps when not in use**. A shelf above the bench is a good choice, as is a shallow drawer (preferably with an organizer insert to prevent sharps getting lost under other materials). Never leave a blade or needle flat on a bench top where someone could not notice its presence or could try to pick it up bare-handed. A Petri dish, jar top, or other small tray-like object serves well to contain loose sharps.

Do not carry uncapped or unguarded sharps across the lab by hand or in the pocket (of a lab coat, perhaps), as another researcher could easily bump into you and either be stabbed or cause you to be stabbed.

When performing higher-risk work, for example, cutting small samples where your fingers must be near the workpiece, **wear cut-resistant gloves** such as woven Kevlar®(DuPont) or chainmail. For fine work such as sectioning tissue, use thinner gloves (the author prefers thin cotton gloves with steel fibers woven through—these are sufficiently thin to permit wearing disposable latex or nitrile gloves over top while still retaining dexterity).

6.2.3 Sharps disposal

Sharps disposal can be the most hazardous part of work with sharps.

Dispose sharps (or clean and stow in their designated storage area) **immediately** after use. Leaving sharps for later disposal creates a high risk of injury—to you or to one of your co-workers.

Use a container—specially-designed **sharps boxes** are ideal, although any container resistant to puncture will do—for disposal of sharps. Fill sharps containers only 75–80% full before sealing and replacing. Sharps containers are **puncture-resistant, not puncture-proof**. Attempting to force a cap on an overfull sharps box can cause sharps to penetrate the container.

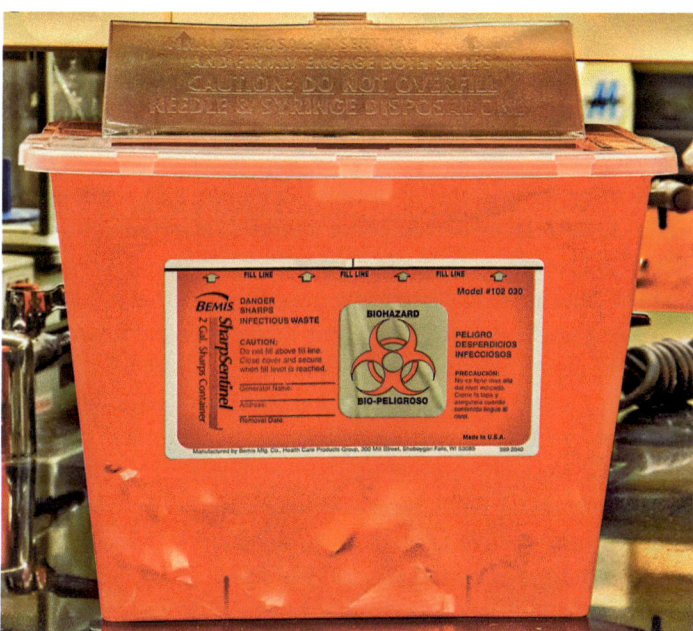

Figure 6.12: Biohazardous sharps box.

Never dispose of sharps in plastic bags. Figure 6.13 shows an improperly-disposed needle that penetrated a custodial worker's hand.

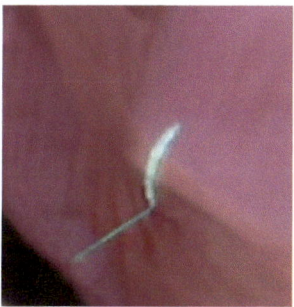

Figure 6.13: Needle sticking through side of biohazard waste bag. Photo courtesy Perry Cooper.

For broken glass, **use a heavy cardboard box** or (if the glassware is contaminated with chemical or biological material requiring containment, treatment, or sterilization) a sealable plastic tub.

Always clean up broken glass with tongs or brush and dustpan, because it is very easy to be sliced while picking up glass shards by hand.

Follow local regulations for ultimate disposal of sharps. At the author's institution, contained sharps are autoclaved or otherwise sterilized (if they are biologically contaminated) and disposed with the laboratory trash (which is either incinerated or autoclaved again & landfilled depending on the application). If the sharps are radioactive from use with radioisotopes, they are disposed as radioactive waste.

6.3 Heat

Elevated temperatures are ubiquitous in the laboratory—scientific and engineering experimentation frequently studies the effect of adding energy to a system, and heat is an easily-available (although low-value) form of energy.

The obvious risk from heat is burns—usually burns to the skin, since that is the usual point of body contact with hazardous temperatures. Burns to the eyes and deep tissue burns are also possible with very strong heat sources, but they are uncommon in the lab.

The other main risk from lab heat sources is fire. It is essential to follow good fire safety practices in laboratories, particularly those which use heat sources or open flames, or those which conduct "hot work" operations (see section 6.3.1 below).

Remember that heat can be transferred by three mechanisms:
- **Conduction** is heat transfer by direct contact with hot material—accidentally touching a hot ceramic piece fresh from the kiln, for example;
- **Convection** is heat transfer by motion of air or other heated fluid—a researcher's lab coat sleeve accidentally held a a few inches above a lit Bunsen burner may ignite even without direct flame contact;
- **Radiation** is heat transfer through space by electromagnetic waves, typically in the infrared (IR) region—a red-hot torch tip set too close to combustible materials may lead to a fire through this mechanism.

6.3.1 Common laboratory sources of heat

Common heat sources used in research laboratories range from the obvious (Bunsen and Fischer burners) to the exotic (electric arcs and solid-phase synthesis reactions). Some of the more frequent heat sources include:
- **Burners**—as mentioned above, Bunsen and Fischer burners are used for direct heating, but torches (such as oxyacetylene torches used for welding or sealing ampoules) are also common in lab;

- **Heat guns**—these tools can reach temperatures of 900 K (650°C; 1200°F), and they retain heat for several minutes when turned off;
- **Furnaces**—materials synthesis and processing make heavy use of high-temperature furnaces;
- **Heating mantles and hot plates**—chemistry labs, in particular, use many of these, but other laboratories also employ them for such purposes as melting wax or heating coatings;
- **Laboratory steam**—whether manufactured in the lab by a steam generator, emitted as a by-product from some apparatus, or piped in from a building steam system;
- **Hot water sources**—ranging from the water tap to exhausts from heat exchangers used to cool another fluid or condense vapors;
- **Cutting, grinding, soldering, welding, and brazing**—mechanical and materials work often requires cutting or joining of pieces, and all of these "hot work" operations use or generate heat;
- **Exothermic chemical reactions**—most spontaneous chemical reactions generate heat, and some reactions can produce a great deal of it;
- **Pyrophoric chemicals**—these compounds, which ignite spontaneously in air, are covered in section 7.4.2.7 below;
- **Lasers and infrared light sources**—we will discuss these sources in detail in chapter 9 below.

6.3.2 Heat-protective apparel

Personal protective equipment for heat comes in two types: PPE for direct handling of hot materials and items intended to prevent the user from catching fire.

Whenever possible, **allow hot samples to cool to room temperature before handling**; this is the simplest method of protection. When the experimental protocol dictates hot sample handling or transfer, use tongs, a spatula, or a heat-resistant sample tray to contain the hot material. Below about 250°C, silicone hot pads or oven mitts work well for short-term direct contact. Above that temperature, specialized protective gloves are preferred for contact applications—metallized gloves of heat-resistant material such as fiberglas ("foundry gloves") can withstand brief contact with molten metals.

Avoid using asbestos gloves found in older laboratories. Asbestos, including the white crysotile variety often used in such gloves, causes lung cancer, and hand operations can easily cause abrasion to the glove, permitting asbestos fibers to become airborne and inhalable.

For extreme high-temperature work, reflective metallized suits, gloves, face masks, boots, etc. are preferred. The reflective nature of such equipment reduces radiant heat transfer between the work area and the researcher. They are frequently heavy and uncomfortable, and they impede body cooling. Researchers who work in such suits should contact an industrial hygienist to determine how long they can remain at work without a "cooling-off" break. Acceptable work times can often be measured in minutes unless auxiliary cooling is used.

For protection against fire in the lab, the PPE of choice is a fire-retardant laboratory coat. Note that **"fire-retardant" does not mean "fire-proof!"** A fire-retardant coat will burn when a flame is applied to it—if the sleeve of the coat dangles above a lit burner, or if the coat is splashed with burning solvent. Nevertheless, when the source of ignition is removed, the coat's fabric will stop burning—it will **self-extinguish**.

Fire-retardant apparel can be made of cotton which has been given a surface treatment or from fire-retardant polyaramid fibers such as Nomex IIIA® (DuPont). **Polyester and other non-fire-retardant synthetic fabrics are not recommended**, as they melt easily and add fuel to a fire. Researchers should avoid cotton/polyester blend laboratory coats if they are exposed to substantial fire risk. (The acrylic sweater that Sheri Sangji wore on the day of her accident—see section 1.1.6 above—is thought to have contributed to her injuries substantially.)

Note that treated cotton lab coats retain fire-retardancy only if properly washed—the surface treatment can be destroyed by some laundering methods, especially those which use bleach. Nomex and similar fabrics are inherently fire-retardant and can be washed without special methods. They are also more durable than cotton and may be cheaper in the long run despite their higher price.

6.3.3 Using torches, burners, and other open flames in the lab

Follow these tips to reduce the probability of a lab fire when working with open flames, heat guns, etc.
- **Remove all flammable and combustible materials** from a 2 m (6 ft) radius around the work area; this includes above and below, not just side-to-side. If you are working in a chemical fume hood or similar location, clear the hood of flammable solvents, combustible laboratory wipes and absorbent pads, etc. Combustible materials that cannot be moved (e.g., wooden shelves or counters) must be shielded with heat-resistant materials.
- **Ensure that no flammable solvents are in use anywhere else in the lab.** In some cases, open solvent beakers or bottles have been ignited by open flames 3–5 m (10–20 ft) from the point of use.
- **Verify that clothing, jewelry, and hair are not loose.** A fire-retardant lab coat is recommended.
- **Know the location of the nearest fire extinguisher or fire blanket**, as well as that of the closest fire alarm pull station.
- **Have all materials needed for the work ready at hand.** None of these materials should be flammable or combustible.
- **Check the equipment that will be producing the heat.** Torches and burners should be in good working order, and you should know how to use them properly. You should also know how to shut them off quickly in an emergency.

- **Ensure you have a safe, stable place to set down hot torches/heat guns/other hot apparatus safely** when you are finished. Setting a hot torch near a box of lab wipes invites a fire.
- **Practice evacuating from the area**. Many people panic when a fire occurs, and panicked people tend to do the first thing which occurs to them, something which may or may not be safe to do. Practice increases the chance that the "first thought" is of a safe escape.

Depending on the work and the work site, hot work operations can be very dangerous. Careless preparation or execution can cause a fire that will destroy the entire building, endanger its occupants, and ruin hundreds of man-years of research. Consult a fire-safety professional (fire marshal, fire safety specialist, insurance technical assistance personnel, etc.) if you have any of the following situations:
- Using open flames or electric arcs near oxidizing, combustible or flammable materials, **including those in sealed tanks**—it is possible to light flammables stored in tanks simply by welding on or near the outside of the tank (and it is possible to ignite steel oxygen tanks themselves by welding on them);
- Heating objects that penetrate walls, such as pipes or tubing—the heat may be conducted far enough to ignite combustibles in the wall or on the other side;
- Holes in floor, ceiling, or wall through which sparks might fly—the sparks can ignite fires on other floors, where you may not see the fire;
- Heating compressed gas cylinders or other pressurized systems, **whether they are under pressure or not**—heating a compressed system may lead to an explosion, and any hot work done on a system designed to contain pressure may reduce the pressure the system can ultimately withstand by detempering or otherwise weakening the materials of construction;
- Any work which in your judgment carries a high possibility of fire.

The fire safety professional can draw on his or her experience and on references such as (US) National Fire Protection Association standards [45] to prescribe additional safety measures. Addition of one or more trained "fire watchers" is a common additional precaution.

6.4 Cold (including cryogen safety)

Cold temperatures are as useful as elevated ones in the lab. The behavior of matter at low temperature is of extreme interest in many fields, and many chemical reactions occur more controllably and with better selectivity and yield when conducted at low temperature.

Depending on the temperature range of interest, cooling may be achieved: by use of refrigeration systems and secondary coolant fluids such as propylene glycol; with dry ice, alone or in acetone bath; using machinery such as a vortex tube (see below); or through the use of **cryogenic** liquids—ultracold gases liquified by refrigeration. Heat transfer from the sample or equipment to the means of refrigeration can occur by all three typical mechanisms: conduction, convection, or radiation. In cooling processes, heat is being removed from the cooled item—cooling systems remove heat rather than "creating cold."

Like hot temperatures, exposure to cold temperatures causes burns. In this case, the burns are **frostburns** (or for lesser burns, **frostbite**). Frostbite injuries differ from heat burns in that the numbing effect of cold makes it difficult to notice that an injury has occurred—and it makes it likely that the injury will be made worse by continued exposure to the cold.

6.4.1 Common laboratory sources of low temperatures

Common laboratory sources of cold temperatures include a variety of equipment and materials.
- **Laboratory cooling baths** cool a working heat transfer fluid, often water, propylene glycol solution, or a silicone oil. Samples may be immersed in the bath for cooling, or the coolant may be pumped to point of use.
- **Vortex tubes** are passive mechanical devices which separate compressed air streams, segregating the hotter molecules from the colder. (These devices are not "Maxwell's demons;" the pressure drop across the apparatus accounts for the entropy loss in the process.) While inefficient and noisy, they are sometimes used for cooling electronic equipment.
- **Laboratory refrigerators and freezers** cover a wide range of temperatures, from just below ambient to about 77 K (–196°C/–321°F).
- **Building chilled water lines** are available in some research buildings, particularly those with an engineering focus.
- **Ice and ice baths** provide a convenient source of moderately low temperatures. If slightly lower ranges are desired, ice-salt, ice-glycol, and ice-alcohol baths are also useful. When using combustible or flammable solvents with an ice bath, apply all necessary fire prevention measures.

6.4.2 Safe procedures for maintenance of refrigerators and freezers

Laboratory refrigerators and freezers usually serve more than one researcher. In such a common-use environment, it is essential both for safety and for the integrity of the experimentation being conducted that refrigerators/freezers and their contents be carefully controlled and maintained. Follow these rules for successful use of lab cooling appliances:
- Keep an inventory of all materials placed in and withdrawn from the refrigerator, **including the owner's name**. When a researcher ceases working at the lab, it is often necessary to recover or destroy his or her research samples. An accurate inventory permits easy identification of exactly what materials are in the refrigerator or freezer.

- Similarly, **clearly identify** all materials placed in the refrigerator or freezer using labels that will adhere under low-temperature conditions. A label which falls off in the freezer is of no use to anyone in the lab—indeed, it is a liability. (The author's institution, which normally provides waste disposal for free, charges approximately €450 ($500) for any unknown sample presented for disposal due to costs for identification and classification.) Be certain to use a marker with ink designed for low-temperature use and **not soluble in the material or solvent in the container**. (Soluble ink inevitably runs and becomes illegible if material drips down the side of the container while making a withdrawal.)
- **Seal stored materials tightly** to avoid leaks and product loss in the refrigerator or freezer. It is nearly impossible to decontaminate many kinds of refrigerator or freezer completely, so it is critical to avoid leaks in the first place. Use close-fitting caps wrapped with paraffin film, or double-sleeve septa—and note that the outer sleeve of such a septum should be pulled down over the glass joint it is sealing. Chemicals should be stored long-term in ampoules or vials with Teflon (or other appropriate material) seals, not in round-bottom reaction flasks with septa. Most septum material degrades rapidly under solvent or product contact, leading to leaks and contamination of the refrigerator.
- **Clean and defrost freezers regularly**. Samples can become trapped in frost layers if defrosting is delayed too long, and the samples may be damaged during the defrost process if they cannot be removed. Energy efficiency of the freezer is also significantly reduced by heavy frost layers. Use foam coolers and dry ice to temporarily store samples during defrost operations.

6.4.3 Cryogenic temperatures

Below the −40°C (−40°F) typically achievable with mechanical refrigeration or ice-salt baths lie the **cryogenic temperatures**. Working with these temperatures involves special materials and methods.

6.4.3.1 Dry ice

The simplest cryogen in common use is **dry ice**, or solid CO_2. Dry ice temperature is −78°C (−108°F). Frost burns are easily possible if dry ice is handled with bare hands; use heavy leather gloves or special cryogen-handling gloves (see section 6.4.3.4 below).

Carbon dioxide sublimates rather than melts at atmospheric pressure, emitting large amounts of CO_2 vapor. This creates the risk of asphyxiation, oxygen deprivation, and CO_2 toxicity when dry ice is used in confined spaces—so use it only in well-ventilated areas.

6.4.3.2 Gases liquified by refrigeration

Several permanent gases are available to lab users in liquid form—a "permanent gas" such as nitrogen cannot be liquified by pressurization, but generally it can be condensed by refrigeration. Such materials are called **cryogenic liquids**. Nitrogen and helium are the most common cryogens used in laboratories.

Liquid nitrogen

Liquid nitrogen (LN) has an atmospheric boiling temperature of 77 K (–196°C/–321°F), and it is typically available in large, insulated Dewar flasks. (See Figure 6.14.) The valving on the flask is usually arranged so that liquid nitrogen or nitrogen vapor can be withdrawn, and the valve arrangement can vary with the vendor supplying the flask. The user typically transfers the required amount of LN into a smaller Dewar flask for use in applications like chilling cold fingers on a Schlenk apparatus or cooling metal samples for low-temperature materials tests.

Figure 6.14: Liquid nitrogen Dewar cylinder.

The 77 K temperature is sufficient to freeze most objects solid in seconds—a classic demonstration made for the public and students is to dip a banana in LN and use it to hammer a nail into a wooden board. Striking the banana too hard usually results in it shattering. The same can happen to flesh, so it is essential to prevent contact between exposed flesh and LN. Appropriate personal protective equipment (see section 6.4.3.4 below) should be worn for dispensing LN. **Phase separators**—sintered metal plugs which help to separate vapor from liquid, preventing spitting and splashing, should be used on the end of LN delivery hoses open to atmosphere.

One particular hazard of LN use is the potential for condensation of liquid air. Oxygen has a slightly higher boiling point than LN (90 K/−193°C/−297°F). When LN is exposed to air, such as in an open-top Dewar flask, oxygen gradually condenses in the nitrogen, forming liquid air. Liquid oxygen (or liquid air) is a powerful oxidizer (see section 7.4.2.6 below), powerful enough to cause explosions upon contact with combustible materials—even small amounts such as grease used on ground glass joints or oils left on the exterior of materials from human handling.

Never "top off" liquid nitrogen Dewar flasks—allow them to evaporate and refill. This will minimize the risk of condensing oxygen in the flask.

LN also expands approximately 700 times upon vaporization, leading to two hazards.

The expansion gas can rapidly exclude oxygen from a workspace, causing loss of consciousness and death by asphyxiation.

Never use liquid nitrogen or other cryogens in a closed or poorly-ventilated area.

The expansion gas can also pressurize a closed container to extremely high levels (millions of Pa/thousands of psi). Unless designed to contain such high pressures, an explosive rupture of the container is likely. Caps supplied with Dewar flasks typically do not seal—they will lift in response to internal pressure, resettling to prevent ingress of air. See section 6.5.1.2 below for more detail.

Liquid helium

Liquid helium (LHe), usually found as mixed isotopes or as ^{4}He, is much colder than LN: the atmospheric boiling point is about 4.25K (−269°C/−452°F). Common uses of LHe in the laboratory include superconductive materials research and cooling for superconducting magnets such as those used in modern NMR spectrometers.

Hazards and precautions given for LN apply equally or more stringently to LHe. The liquid-gas expansion ratio for LHe is about 1:750, compounding the asphyxiation and pressurization hazards, while the potential for condensation of oxygen is also heightened [46].

LHe cylinders (see Figure 6.15) are more complex than those used for LN, and special training should be obtained. In particular, LHe flasks are prone to development of ice plugs if improperly operated; if moist air is admitted to the flask's internal piping, the moisture will freeze. These plugs act to seal the Dewar flask, preventing release of normal boiloff vapor and creating the risk of a pressure explosion.

Special techniques are required to remediate this situation, and any researchers who work with LHe Dewar flasks under conditions where this may occur should be equipped with the appropriate equipment, procedures, and training to remove such plugs before a serious risk of explosion develops. Contact your supplier for these materials and training.

Figure 6.15: Liquid helium Dewar cylinder. Photo courtesy Boris Steinberg.

Other cryogenic gases
Other cryogenic gases used in the lab include argon, oxygen, and carbon dioxide, all of which can be obtained in liquid form. Argon and carbon dioxide carry similar hazards to LN. Liquid oxygen's (LOX) oxidizing properties dictate that researchers should obtain extensive training before working with it, and the amount used should be minimized. Specially-cleaned and -qualified equipment is necessary for work with LOX. Explosions involving oxygen can be quite spectacular and deadly—one case in Africa resulted in the unrecognizable remains of one victim being presented to the coroner in a small bucket [47]. (See Figure 6.16.)

6.4.3.3 Liquified pressurized gases
It is possible to liquify many gases by application of pressure. Cylinders of these materials are sometimes partway full of liquid. Carbon dioxide and propane are the most common examples of liquified pressurized gases, but anhydrous ammonia, nitrogen dioxide, and others are available.

Cylinders of these gases are typically supplied in a configuration designed to deliver vapor—the liquid present only acts to increase the cylinder capacity. If it is necessary to deliver liquid from a cylinder, a cylinder equipped with an **eductor tube,** or "dip tube," may be ordered from the supplier. Cylinders should **never** be turned on their sides or inverted in an attempt to withdraw liquid (unless they are specifically designed for this use).

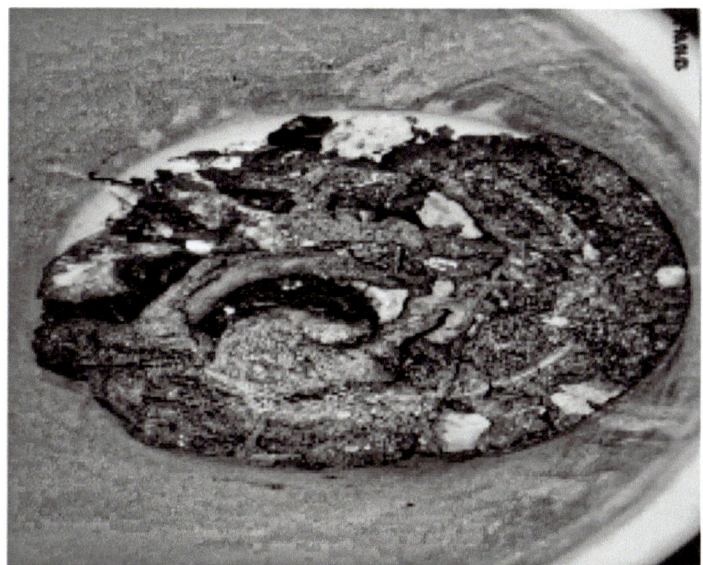

Figure 6.16: Remains of worker killed in oxygen cylinder explosion [47].

Pressure safety relief valves and rupture discs (see section 6.5.1.2 below) are usually designed to relieve vapor. Any closed system using liquified pressurized gases should employ pressure safety relief designed for liquid and vapor-liquid two-phase flow; consult a qualified mechanical engineer or other designer for a proper design.

6.4.3.4 Cryogenic protective apparel

Because handling of cryogens usually involves close proximity to the material, personal protective equipment is particularly important.

Wear long sleeves and long pants, neither of which should be rolled up or have rolled cuffs—rolls in the clothing tend to trap cryogens next to the skin in the event of an exposure, making the resulting frostburn much worse.

Wear eye protection at all times when handling cryogens. Chemical splash goggles are recommended for laboratory use, and for any quantity above a few mL, a face shield is also suggested.

Wear heavy leather gloves for short-term handling of dry ice. Longer-term handling of solid CO_2, or any handling of LN/LHe or other ultra cold liquids, requires special cryogenic protective gloves (see Figure 6.17.) Note that gloves constructed for hot temperatures, such as woven autoclave gloves (see Figure 6.18), should **never** be used for cryogens. Insulation intended for high temperatures is not necessarily effective against low temperatures, and loosely woven material presents little protection against liquid cryogens.

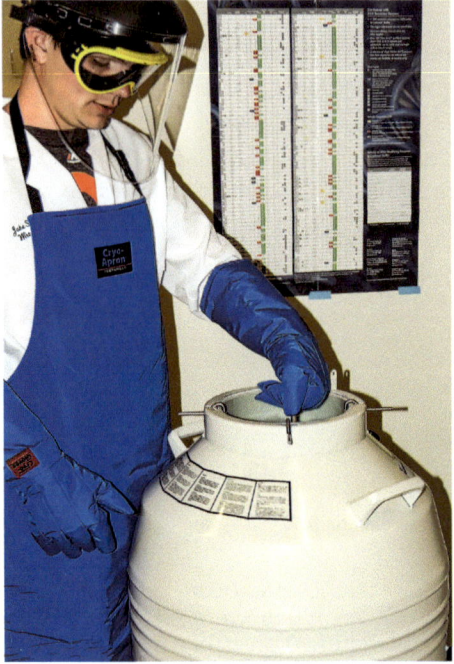

Figure 6.17: Researcher using cryogenic PPE.

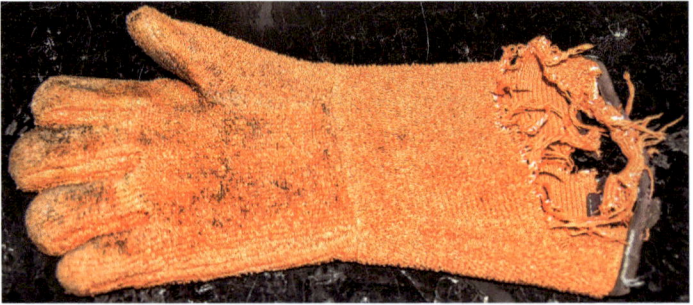

Figure 6.18: Autoclave gloves.

For handling of larger (liter-scale) amounts of cryogens, such as when withdrawing from a large LN Dewar cylinder, **wear a cryogenic apron** (see Figure 6.17). These are constructed similarly to cryogenic gloves, and they provide protection against spills or splashes to the front of the body.

6.5 Pressure and vacuum

Pressure and vacuum sources are ubiquitous in the research laboratory. From "house air," available on tap at the bench, to compressed gas cylinders containing a variety of hazardous materials, pressure can apply a great deal of force, particularly to objects with large surface areas (such as the interior surface of a compressed gas cylinder). In this section, we will deal primarily with hazards from compressed gases and vacuum sources.

6.5.1 Compressed gases

After hazardous chemicals, compressed gases are probably the second most common type of hazard encountered in the lab. Sources of compressed gases include prepackaged cylinders, building air compressors, and boiloff vapor from cryogen use.

6.5.1.1 Laboratory air

Many laboratories employ compressed air for such purposes as operating air-actuated mechanical devices, providing cooling to equipment, operating sandblasters or airbrushes, or aerating liquids. Some laboratory buildings are equipped with building air compressors and air-distribution piping, while others use compressed air cylinders for the same purpose. In permanent installations, there are air taps installed on the benches or walls.

The typical building air system is pressurized in the range of 400–800 kPa (approx. 70–120 psi). This pressure is sufficiently high to propel parts, dust, metal chips, etc. around the lab with great force and supersonic speed. Care is required when using compressed air to blow off debris, and goggles of a type appropriate to the material being blown are necessary. In the case of smaller particles, which are respirable, respiratory protection may also be necessary. Note that the air velocities involved are sufficient to overwhelm the capture abilities of a chemical fume hood or biological safety cabinet, and so use of compressed air in such areas can spread contamination throughout the lab and is not recommended.

> *Pressure-injection hazards from compressed air*
> It is an unfortunately common practice in some laboratories and shops to use building compressed air systems to clean **people**, either directly by aiming the stream to blow off dust and dirt on one's body or indirectly by holding a part (such as a pipette rack being blown dry) with bare hands and directing the airstream too close to the hands. Horseplay with compressed air guns is a common source of injury as well.
>
> This practice introduces three serious hazards [48]: damage from particles propelled by the air, hearing damage, and injury (and potential death) from air injection under the skin. For this reason, cleaning humans with compressed gases or deliberately directing a gas stream at a person is illegal in many jurisdictions and unwise in general.
>
> Cleaning objects with compressed air necessarily results in high-velocity airborne particles. These particles may penetrate the skin or eyes, abrade skin, damage other objects, or be inhaled. Depending on the properties of the materials that have become airborne, damage may be light (minor abrasions) or severe (permanent eye injury, lung disease).
>
> Compressed air streams may be extremely loud, and persons using them for cleaning are in close proximity. Hearing damage has occurred from exposure to such noise levels (see section 6.7.3 below).
>
> Finally, and most seriously, air pressures of 200–305 kPa (30–50 psi) are sufficient to drive air or other compressed gases under the skin, inflating the affected tissue. This is said to be excruciatingly painful, and amputation of the affected member is necessary in many cases. It can also be deadly, as small bubbles can enter the bloodstream, causing air embolisms (blocked blood vessels) in critical areas such as the brain—effectively causing a stroke or heart attack [49]. In controlled situations, this air-injection effect is used as a substitute for hypodermic syringe injection [50].
>
> **Never** direct a stream of compressed gas at a person, particularly at exposed skin. If the possibility exists of inadvertent exposure, such as with compressed air used to clean parts in a machine shop, always use air nozzles (air guns) designed to produce no more than 130–200 kPa (20–30 psi), **even when pressed directly against the skin.** Such air guns are often sold as "safety nozzles," but using one is insurance against disaster, not permission to aim compressed air at someone.

6.5.1.2 Pressure and vacuum safety relief

A pressurized gas must be contained in order to prevent expansion to atmospheric pressure; if the pressure is inadvertently or adventitiously created, provision for expansion must be made. If the pressure inside a system exceeds the ultimate strength of the materials from which the system is constructed, the containment will fail in a **pressure explosion**. Such explosions generally convert the stored energy of the pressurized gas into kinetic energy in the form of blast or high-speed shrapnel.

In order to prevent pressure explosions, all pressurized systems, whether they are intended to contain pressure or if they are merely capable of becoming pressurized during normal or abnormal operation, **must be protected from overpressure**. There are three basic methods for providing this protection.

System design. The system may be designed such that it is capable of withstanding the maximum pressure possible during operation or emergency situations. This approach is often used with consumer equipment such as refrigerators or air conditioners. When the pressurized gas is dangerous (e.g., highly toxic or chemically explosive), appropriate system design can also be a useful technique. Careful analysis and engineering is required to ensure that suitable design parameters are used. It is recommended to consult a qualified professional engineer for such analysis—even if you have a degree in mechanical or chemical engineering.

Rupture discs. A part of the system may be designed to fail at a predetermined pressure, lower than the design pressure of the rest of the system—thus, that portion would be expected to fail first. If the resulting opening into the system is sufficiently large, extra pressure is released in a controlled manner and catastrophic over-pressurization of the system can be prevented.

Pressure safety relief valves (PSRVs). Spring-loaded or other automatically-operated valves may be incorporated into the system, set to open below the system design pressure. Similar to a rupture disc, a PSRV permits controlled release of pressurized material and prevents over-pressurization. The advantage of PSRVs over rupture discs is that they are generally designed to close once the pressure drops to a predetermined level, minimizing the loss of material from the system. (Whether PSRVs actually close after operation depends on a number of factors such as the quality of the valve, the amount of maintenance it receives, and the presence of solids in the released material. The author generally assumes that any PSRV that has lifted will at least leak after closing.) Laboratory-sized PSRVs are generally available from scientific fitting suppliers.

While pressure-resistant system design depends on the designer's ability to properly determine the maximum possible system pressure, the rupture disc and the PSRV approaches depend on several factors to operate properly.

- The flow capacity of the PRSV or of the "hole in the system" created by a burst rupture disc must be sufficient to release pressure at a rate at or above the rate at which pressure can build up in the system. The rupture disc on a compressed gas cylinder in a room on fire, for example, must release vapor at least as fast as the fire is creating it inside the cylinder (through evaporation of compressed liquid or expansion of contained gas), lest the cylinder explode. Flow through a relief device is generally limited by the speed of sound in the relieved material—sonic flow "chokes" the relief rate to a certain maximum.
- The relief system (relief device and associated piping/tubing) must be able to handle all material that may flow into it. In some cases, this may include things other than gas or vapor, such as liquids or even solid material.
- Pressure relief for systems that contain chemical reactions are particularly difficult to design and require personnel trained in special methods [51], since the number of phases and number of molecules in the system are changing even during the relief flow. Polymerization reactions are particularly difficult to relieve since the relief device may become clogged with polymer forming within it.
- If released material is dangerous (which it generally is—even steam relief can be dangerous due to the high temperature), consideration must be made for

disposal of the relieved material. This may be as simple as routing the relief output into a chemical fume hood exhaust plenum or as complicated as installing specially-designed treatment systems. The latter is required for relief of toxic materials in some areas. Note that long runs or many direction changes in a relief outlet line reduce the capacity of the relief system—yet another consideration for the designer.
- Lastly, all relief systems require regular maintenance if they are to be relied upon to work properly. Unmaintained PSRVs may not open, rupture discs may be blocked, system working pressures may be reduced by corrosion, and the relief system outlet line may become blocked by water, ice, rodents, insect nests, etc.

Laboratory users of pressurized systems are **strongly** advised to consult with a qualified mechanical or chemical engineer for assistance in designing appropriate pressure safety relief for pressurized systems.

6.5.1.3 Compressed gas cylinders

Only certain gases are available "on tap" in typical labs: air, nitrogen, natural gas, and sometimes argon and oxygen. Of these, only air, oxygen, and nitrogen can normally be supplied from fixed equipment drawing directly from the atmosphere, such as air compressors or nitrogen/oxygen separation systems. Most lab gases, including those mentioned above, are obtained from suppliers in compressed gas cylinders, heavy-wall tanks constructed to hold gases under high pressures and gases liquified by pressure or refrigeration.

Asphyxiation and other hazards of compressed gases

Using compressed gases exposes researchers to several different types of hazards, the most obvious of which is sudden pressure release. Most lab workers harbor suspicions that gas cylinders are "bombs" capable of exploding at any time. This is not so. Compressed gas cylinders are heavily engineered to withstand the rigors of rough transportation, handling, and use in adverse environments—it takes effort or severe neglect to make a cylinder explode.

The pressurized systems to which cylinders are connected, though, **are easily capable of throwing off parts at high velocity or even of exploding**. Typically, this happens as a result of over-pressurizing the system, either by mistake or by poor design. A simple lab example of the latter is connecting a regulator that can deliver 3.5 MPa (500 psi) to a system rated for 200 kPa (30 psi), especially if the system is not designed with adequate pressure safety relief.

Pressure leaks from pressurized systems can form high-pressure gas jets capable of gas injection beneath the skin (see section 6.5.1.1 above) or even cutting materials (including severing or damaging body parts).

Aside from certain grades of compressed air and special breathing-gas mixtures (e.g., helium-oxygen mixes used for deep diving), the contents of compressed gas cylinders are not breathable. Asphyxiation is a constant hazard when working with compressed gas cylinders. Typically, oxygen levels below 19.5% are considered oxygen-deprived. Note that the brain mechanisms that determine when to breathe sense carbon dioxide levels in the blood, not oxygen levels—this means that many unbreathable atmospheres **are not detectable through shortness of breath**. It is perfectly possible for a leak to slowly decrease the oxygen concentration in a laboratory, unnoticed until researchers begin to pass out. Installing oxygen sensors in laboratories where large leaks of compressed gases or cryogens are possible should be evaluated.

Chemical hazards of the compressed material are also a concern; see chapter 7 below for more detail. Flammable gases create fire hazards in the lab; control of ignition sources and effective laboratory ventilation is key. (See section 2.3 above.) Flammability properties of gases can be significantly altered by pressurization; multiple industrial explosions of mixtures of air with chlorodifluoromethane (Refrigerant—22; R—22), for example, have occurred even though R—22 is considered "nonflammable." Oxidizing gases such as oxygen and nitric oxide accelerate combustion drastically, easily igniting materials such as steel, aluminum, and even those such as poly(tetrafluoroethylene) (PTFE; Teflon®—DuPont) not normally thought of as combustible. Oxidizer-fed fires can be difficult or impossible to control. **Even compressed air** can act as an oxidizer, igniting tubing and fittings in the presence of a strong enough spark. Other chemical properties—corrosivity, self-polymerization hazards, etc.—also pose risks.

Response to laboratory emergencies

Despite all efforts at prevention, emergencies are still possible in a laboratory. Observe the general principles below:
- First and foremost, you should **know your facility or institution's general emergency procedures**—for example, are you expected to attempt to extinguish small (incipient) fires or are you expected simply to evacuate? (The author strongly recommends evacuating and leaving the firefighting to professionals.) Also memorize the local emergency number, whether that is a public emergency communications center or a local emergency response team hotline.
- **Anticipate possible emergencies** that may occur in your lab. If your laboratory uses highly flammable solvents, for example, a solvent fire is a distinct possibility. Make a plan for such foreseeable events, and ensure that each researcher has memorized it.
- **Practice** emergency procedures, particularly evacuation. The author has experienced laboratory fires and fires in the home. In both cases, some of those present panicked and did the first thing that occurred to them. Practicing an evacuation procedure helps to ensure that that "first thing that comes to mind" is something safe.
- **Pre-position** needed emergency supplies in the lab, somewhere they will be accessible in an emergency. If your lab uses large amounts of acid, it would be prudent to keep a large bag

of sodium bicarbonate where it can be reached in the event of a spill. (Sodium bicarbonate, being an amphoteric—both acidic and basic—substance, is an effective neutralizer for many types of corrosives. Check for compatibility with the particular compound you wish to neutralize first, though, to avoid hazardous reactions.) If it may be necessary for you to fight your way out of a burning laboratory, positioning a fire extinguisher where you can reach it if you become trapped would be wise.
- **Do not rush in to assist** if you see someone unconscious in a laboratory, unless you know **exactly why** that person is down. You may add yourself to the list of victims. What if, for example, the room's atmosphere is unbreathable? You will rapidly lose consciousness as well, and no one will know to come and rescue you. Call for professional rescue.

Anatomy of a gas cylinder

Compressed gas cylinders have many individual parts, each with a specific function. It is **essential** for cylinder users to recognize and understand how each of these parts works to maintain a safe gas delivery system. Figure 6.19 shows an overall view of a typical compressed gas cylinder holding about 8.5 m³ (300 SCF; standard cubic feet) of gas. Important elements of the cylinder system include:

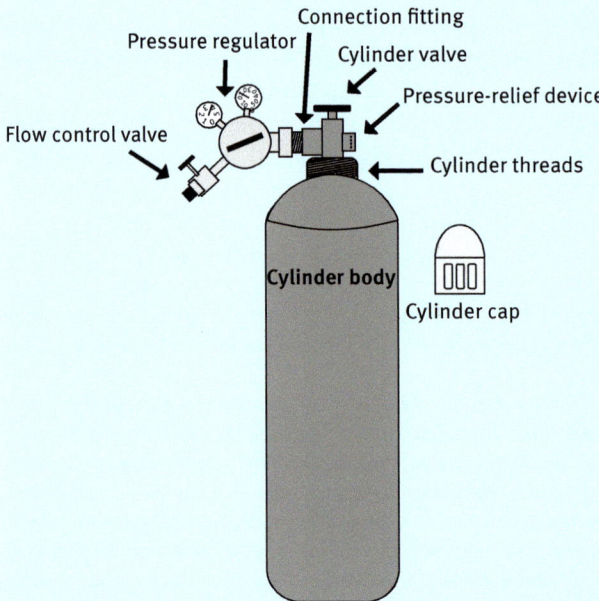

Figure 6.19: Compressed gas cylinder, with major parts labeled.

- The **cylinder body**. In order to withstand heating from solar load in open transport trucks, high temperature in hot storage rooms, etc., this heavy-wall pressure vessel is typically constructed to withstand several hundred to several thousand psi (several MPa) additional pressure than that exerted by the intended contents at room temperature. Cylinder bodies are

usually steel, but aluminum and exotic alloys are used for ultra-pure or chemically-aggressive gases. In the case of compressed air cylinders intended to be carried as part of a self-contained breathing apparatus, the body may be made of carbon- or fiberglas-reinforced resin.
- The **cylinder valve**. This (usually) hand-operated valve controls flow of gas from the cylinder.
- The **cylinder threads**. This threaded ring, located above the shoulder of the cylinder body, allows attachment of a **cylinder cap** designed to protect the hand valve from damage.
- The **pressure-relief device**. This small rupture disc will open to relieve internal pressure in the event of an overpressure event. This is usually due to fire but can also be due to backflow of gas into the cylinder or an internal chemical reaction such as polymerization or combustion. Pressure-relief in larger cylinders or those used for cryogens such as liquid nitrogen is often accomplished by using a spring-loaded pressure safety relief valve that closes once pressure has been reduced to a safe level.
- The **connection fitting**. Also called a "CGA fitting," after the Compressed Gas Association, which standardizes some types of connections. Some cylinders, particularly those for electronic gases, use "DIN connections," which follow standards set by the Deutsche Institut für Normung (German Institute for Standardization). **Connection fittings are unique to the type of gas in the cylinder.** (Often multiple gases of similar type—inert gas, simple flammable, etc.—share the same type of connection.)
- The **pressure regulator**. This device (connected to, but separate from, the cylinder itself) controls the output pressure delivered from the cylinder, maintaining it relatively constant as the pressure in the cylinder drops with usage. Multiple stages of regulation are available for high-accuracy applications. **Note that a pressure regulator does not control flow.**
- The **flow control valve**. This device, typically a needle valve, allows fine control of the amount of gas being delivered from a cylinder system. **The flow control valve does not control pressure.**

Safe practices with compressed gases
Follow the practices below in order to work more safely with compressed gases.
- **Ensure that equipment cannot be subjected to pressure in excess of its design working pressure.** Any equipment that can, during normal operation, abnormal conditions, or standby situations be over-pressured **must** have adequate pressure safety relief (see section 6.5.1.2 above). A common error is to use typical laboratory tubing, which is usually rated for 150–300 kPa (20–40 psi), with a cylinder regulator capable of delivering 20 MPa (3 kpsi)—a small error in setting the delivery pressure will lead to explosion of the tubing (which is often immediately next to the researcher's face at the time and may strike it like a whip).
- **Minimize the number of cylinders in your lab.** Most lab gases can be delivered on short (2–3 day) notice, excepting ultrapure/rare gases or special mixtures. Many laboratory buildings have no established cylinder storage areas. Excessive cylinder load in the labs is an unnecessary safety risk and may be illegal under local fire codes.
- **Do not retain cylinders more than 5 years**, as they require periodic pressure tests by the vendor. This is economically favorable for the lab, also, as it minimizes monthly **demurrage charges** (rental fees for cylinders).
- **Restrain cylinders at all times.** The only time cylinders should be free of restraint is when they are being transferred from a wall-mounted strap or chain point to a transport cart. Keep the distance over which cylinders must be moved to perform this transfer to an absolute minimum. Do not overload cylinder supports—most are constructed for a single cylinder.

Restraints should surround the cylinder approximately 2/3 of the way up the cylinder—not at the bottom or around the neck.

- **Do not use cylinder transport carts as permanent cylinder supports.** In particular, **never** use a gas cylinder while it is on a cart, unless the cart is specifically designed for that purpose, like certain welding carts.
- **Leave cylinder caps on and regulators unmounted when the cylinder is not in use.** "In use" means that you or a fellow researcher expects to use the gas within a few minutes. The cylinder cap protects the hand valve atop the cylinder—damage to this valve can cause the cylinder to become a high-speed missile capable of reaching 100 km/hr (60 mph) and/or 30 m (100 ft) height. Anyone in the way of such a missile may be killed, and any equipment or walls may be destroyed.
- **Do not turn cylinders on their sides or invert them.** The sides are the weakest portion of the cylinder, and inversion creates the risk that the valve will be damaged. Sometimes researchers invert or turn cylinders in order to deliver liquid from the cylinder instead of vapor (e.g., with carbon dioxide cylinders, which are partway full of liquid CO_2). Proper practice is to order the cylinder with a **dip tube**, or **eductor tube**, which pipes the hand valve to the bottom of the cylinder so liquid is delivered instead of withdrawing from the headspace at the top.
- **Never use a cylinder without a pressure regulator and flow control valve.** The cylinder valve does not effectively control pressure or flow. It mainly functions as an on-off valve. Do not attempt to use the cylinder valve as a pressure or flow regulator; you may lose control of the operation.
- **Use the right regulator.** The regulator supply range should be approximately 25% more than the pressure of a full cylinder, and the required delivery pressure should be in the middle of the available delivery range from the regulator. Many regulators do not deliver reliable pressure control outside 15–85% full range.
- **Never substitute connection fittings on a regulator or use adapters** to convert one type of fitting to another unless you have been specifically trained to do so by the gas supplier.
- **Secure all hose connections with clamps.** Use spring or screw clamps if gas is to be conveyed through tubing connected to a hose barb.
- **Never use an oxygen or other oxidizer regulator with other materials (or vice versa).** The materials of the regulator or residual gases present from other uses may spontaneously and violently combust, rupturing the regulator and turning it into shrapnel.
- **Use non-sparking tools (regulator wrenches, etc.) with compressed gas systems.** As noted above, even compressed air is capable of igniting many metals and other materials. Aluminum-bronze or copper-beryllium alloy tools are most suitable; the aluminum-bronze usually costs less.
- **Always provide adequate pressure relief capacity in a pressurized system.** While cylinders and regulators have their own internal pressure relief, any pressurized system should be either equipped with a relief valve capable of relieving the maximum flow from the cylinder or designed to withstand the full delivery pressure of the regulator. Consult with a qualified mechanical or chemical engineer if you do not know how to properly size pressure safety relief systems. Route pressure relief lines to a safe location.
- **Never block the pressure-relief outlet.** This compromises the safety relief and may create conditions for an explosion to occur.
- **Always test systems for leaks at low pressure before exerting full pressure on the system.** Low-pressure leaks can easily turn into violent high-pressure failures or release large amounts of hazardous materials into the lab atmosphere.
- **Always open the cylinder hand valve as little as possible.** This restricts the flow rate of the resulting leak in the event of a catastrophic failure somewhere in the apparatus. Do not open cylinder valves fully unless the desired flow cannot be achieved with a narrower opening.

6.5.1.4 Cryogenic boiloff

An additional source of hazardous pressure that is often overlooked is boiloff vapor from cryogenic liquids. As mentioned above (see section 6.4.3), all cryogens expand greatly upon heating. Since no storage container is completely insulating, heat does penetrate and create boiloff vapor. If the system is sealed, pressures can become **very high**, easily exceeding the burst pressure of the system.

Case study: Explosion of a sealed Dewar flask

One Friday afternoon in 2009, a physics student withdrew several liters of liquid nitrogen into a medium-sized Dewar flask (mounted on wheels for transport). For unknown reasons, the Dewar flask had been modified with a hand valve so that it could be isolated to prevent losses of expensive cryogen, but no pressure relief device was installed to vent boiloff vapor. The student then departed for the weekend, leaving the sealed flask in the laboratory.

Figure 6.20: Aftermath of laboratory Dewar flask explosion. Photo courtesy Perry Cooper.

The small amounts of heat that penetrated the Dewar flask were sufficient to create internal pressures estimated at approximately 1200 psi (8 MPa). Since the flask was not designed to withstand anything above atmospheric pressure, it burst during the mid-afternoon two days later. (See Figure 6.20.) The lab was seriously damaged, with much of its contents destroyed. Had any researchers been present in the room at the time of the explosion, they could have been seriously injured or killed.

6.5.1.5 Other lab pressure sources
Some labs make use of other sources of high pressure:
- **Compressors**, whether small reciprocating or scroll-type laboratory units or larger screw or centrifugal compressors used in pilot-scale laboratories (such as a refrigeration engineering lab) are capable of producing high pressures—how high depends on the design of the equipment. While such compressors generally have multiple independent high-pressure safety switches that cut power to the equipment in the event of an incipient over-pressure, this protects only the compressor itself, not necessarily the entire pressurized system.
- Ultra-high pressure research (e.g., 300 MPa; 45 kpsi) often employs **manual pressure generators**. This equipment is designed to produce very high pressures, and its use is recommended only for those experienced in high-pressure operations.
- **Gas-generating chemical reactions** also produce pressure. An example would be the use of oxalyl chloride to synthesize acyl chlorides—this reaction produces hydrochloric acid as a by-product. The acid is often emitted from the reaction mixture as bubbles of gas, and if the container is sealed or has insufficient relief capacity, the apparatus can explode. Relief design for reacting systems is a specialist skill [51] only infrequently possessed by academic personnel (including principal investigators)—consult a qualified chemical engineer to ensure that reliefs for sealed reacting systems (particularly polymerization reactions) are sufficient to avoid explosion.

6.5.1.6 Vacuum
Vacuum equipment also carries pressure-related hazards, since a container under vacuum exhibits a pressure differential between it and the outside environment—the pressure is merely exerted on the **outside** of the container instead of the inside. An example accident would be a filter flask that implodes while vacuum is applied, or a sealed pressure tank crushed by the vacuum formed when it is steam-cleaned and then sealed without cooling completely.

Vacuum in laboratory equipment comes from three sources:
- Applied vacuum from a building vacuum system or a local vacuum pump;
- Physically-induced vacuum from the condensation of steam, hot vapors, or perhaps cryogenic boiloff vapor; and
- Chemically-induced vacuum from conducting a reaction that consumes a gaseous reactant, lowering the pressure.

Vacuum failures, particularly implosions of brittle materials such as glass, are notorious for sending clouds of shrapnel across the lab.

The presence of pressure or vacuum equipment in a laboratory is sufficient reason to wear impact eye protection such as safety glasses **at all times.** It is often not possible to predict when pressure or vacuum failures will occur, so it is necessary to protect yourself consistently.

6.5.1.7 Safe practices with vacuum
When using equipment under vacuum, follow these recommendations:

- **Verify that the physical condition of the equipment has not been compromised.** Glass vacuum flasks, such as those used underneath biological safety cabinets, can acquire scratches that weaken the material.
- **Wrap vacuum glassware with strong tape.** This will reduce the amount of shrapnel sent flying if the glassware implodes.
- **Verify that the equipment is rated to take the vacuum that will be applied.** Just because a piece of equipment is designed to handle a soft vacuum of 50,000 Pa (7.3 psia) does **not** mean that it will not implode upon application of a hard 3 Pa (4.4×10^{-4} psia) vacuum.
- **Supply any equipment that may be subjected to vacuum beyond its design limits with vacuum relief.** This may take the form of a vacuum breaker valve or, more simply, a glass bubbler.
- **Break vacuum slowly.** Sudden pressure release stresses the equipment's materials of construction and can reduce the useful life of the equipment (as well as introduce a possible mode of failure).
- **Never allow chemically, biologically, or radiologically hazardous materials to enter a vacuum system.** Use filters, treatment systems, and other equipment as necessary to keep vacuum pumps, building vacuum, etc., free of contamination. If you do not, you are placing the personnel who maintain the vacuum system at serious personal risk of exposure.

6.6 Electricity and magnetism

Even more than pressure and vacuum, **electricity** and **magnetism** are ubiquitous laboratory hazards. Electricity, in particular, tends to be discounted as a laboratory safety risk. This is because it is a utility—we turn on the light switch on the wall or plug equipment into a wall socket, and electricity appears, on our command. The natural expectation is that we can just as easily turn it off to avoid exposure. This is not always true.

6.6.1 Electricity

Electricity is the motion of electrons through a conductive substance, brought about by gradients in electrical potential, that is, voltage. Because materials ranging from the conductor itself to human body parts in contact with it can be damaged by electricity, three major hazards arise: **shock, arc,** and **blast.**

The hazard posed by electricity should not be underestimated or discounted. At the dawn of the electrical age, electrical safety risks were not well-understood. In the 1880's, the death rate for electrical linemen was at least **double** that for other occupations, and of those who were linemen in 1880–1900, **half of all linemen deaths occurred from accidents on the job** [52].

6.6.1.1 Shock

The electrical hazard most researchers think of first is **shock**. Electric shock is the result of current flow through the body. It occurs when two body parts are touching points that are at different voltages. An example would be the common (and often fatal) shock experienced by persons touching electric wires with wet hands. The hand, apart from the effect of surface resistance, is effectively at the voltage of the wire, while the feet are near the local electrical ground. If the voltage is enough to overcome the body's inherent electrical resistance, current flows from wire to hand to foot to ground.

Voltage itself does not cause electrical injury: it is the flow of current that does the damage. Higher voltages mostly lead to surprise and pain. This leads to a saying among radio technicians: "Volts jolt; mils [milliamperes] kill." Very little current is needed to induce severe or fatal injury, particularly with alternating current. Sometimes only 30 mA is needed to stop a victim's breathing, and approximately 100 mA can induce ventricular fibrillation—effectively stopping the heart from operating effectively. Other physical effects of shock can be found in Table 6.1.

Table 6.1: Physical effects of shock

Current (mA)	Effect
< 1	Generally not perceptible
1	Faint tingle
5	Slight shock felt; not painful but disturbing. Average individual can let go. Strong involuntary reactions may lead to other injuries
6–25 (women)	Painful shock; loss of muscular control. Person may be thrown away from power source by contraction of extensor muscles.
9–30 (men)	The freezing current or "let-go" range. Person may be thrown away from power source by contraction of extensor muscles; voluntary breaking contact with power source not possible.

Table 6.1: (continued)

Current (mA)	Effect
50–150	Extreme pain, respiratory arrest, severe muscle contractions, possible death.
1000–4300	Rhythmic pumping action of the heart ceases, severe burns, likely death.
10000	Cardiac arrest, severe burns, probable death.

Source: [53]; cited in [54].

Regrettably, thresholds for these effects are known rather accurately, due to war crimes committed during World War II by Axis medical personnel experimenting upon prisoners of war.

Notably, some shock injuries result only indirectly from the electricity. At 10 mA or above, paralysis or muscle spasms are possible. A common shock-related injury is falling off a ladder or elevated platform as a result of such spasms.

Dry skin has a conductivity of approximately 30 kΩ, while damp or sweaty skin conductivity is about one-tenth of that value. Applying Ohm's Law, we find that 120 V/3 kΩ = 40 mA, easily enough to cause respiratory arrest, explaining the common caution never to touch electrical equipment or switches with wet hands. Such calculations are very rough; under the right conditions, 24 V exposures can be sufficient to kill.

Death from cardiac effects is particularly likely if the current flow path passes through or near the heart. When working with electrical equipment, **always** ensure that the hands touch points at similar or equal voltages—holding a grounded part such as an equipment casing while working on electrified equipment with the other hand routes the current through the chest. When they cannot avoid working on live equipment (such as during troubleshooting operations), many electrical technicians work with one hand in a pocket to prevent inadvertent contact with grounded objects.

6.6.1.2 Arc & blast

When voltages become high enough, any material will conduct electricity, even insulators such as air. This **dielectric breakdown** leads to sudden current flow, often at very high currents. Air gaps are particularly prone to breakdown at high voltage, since the dielectric strength of air is only about 3 MV/m (for comparison, mica and polytetrafluoroethylene—PTFE, or Teflon®—often used as insulators, have strengths on the scale of 100 MV/m). The current flow through air as a result of a dielectric breakdown occurs through a channel of very hot air called an **electric arc.** This air, heated to a plasma, provides a low-resistance pathway through which high currents can flow. Note that the voltage that initiated the arc need not continue at such a high level—the plasma channel is maintained by the current flow, not by the voltage. This means that transient voltage spikes in a normally low-voltage system can create lasting electric arcs.

In addition to dielectric breakdown of air, arcs can form when excessively high currents in a circuit melt conductors or through an electrical resonance called the **inductive flywheel effect**. (An inductive flywheel occurs when a de-energized circuit is located near an alternating current source, which induces current in the unpowered circuit through magnetic coupling; if the induced oscillations occur at a resonant frequency, very high currents and voltages can develop at circuit contacts.)

Arc temperatures can reach approximately 50 kK at the terminals of the arc, causing injury or damage through radiational and convective heat transfer over long distances. (Second-degree burns are possible over 4 m (13 ft) from a strong arc. Eyes are subject to physical injury as well as severe burns from the ultraviolet radiation produced by the hot arc, and the body may receive burns of any degree. Clothing or other combustibles may ignite spontaneously.

In addition to injuring researchers directly, the heat can rapidly and explosively melt nearby parts, such as the casings of electrical equipment, ejecting hot, atomized metal and other materials in a **blast** effect. While the vaporized material is toxic (and likely to be inhaled), blast can also create over-pressures in the air of 5–10 kPa (0.1 atm), sufficient to blow over concrete walls and turn electrical switchgear into shrapnel. A system with a potential for electric arc is **not safe just because it is enclosed**.

6.6.1.3 Good practices for using electrical/electronic lab equipment

Follow the practices below to work more safely with electricity:

- **Inspect all electrical equipment thoroughly** before using. Frayed wires, damaged plugs or sockets, slight scorch marks all can indicate problems that can cause you immediate injury. Electrical tape is not a substitute for replacing a faulty cord. See Figure 6.21 for an example of equipment that should not be used.
- Whenever possible, **use ground-fault circuit interrupters (GFCIs)**. These devices, available as plug-in units or as wired-in sockets (preferred) effectively compare the amount of current being supplied to the equipment with the amount being returned on the neutral wire. Any difference indicates a **ground fault**, meaning that current is being diverted through another path to ground—-potentially through your body. Upon detecting a ground fault, the GFCI shuts off the current. Test GFCIs frequently using a leakage current tester and follow local recommendations for replacing malfunctioning GFCIs.
- GFCIs are sometimes said to protect you from contact with energized conductors by wet body parts (which is why they are often required in bathrooms and other wet areas). **Do not depend on this protection**. GFCIs require finite time to activate, and it is possible to receive a fatal shock before the GFCI shuts off the power. Do not take the presence of a safety device like a GFCI as permission to relax other safe practices, such as drying your hands completely before handling electrical equipment.

Figure 6.21: Damaged heat gun. Note fraying at several locations along the cord.

- **Do not use adapters** to connect three-prong electrical plugs to two-prong outlets. These "cheater plugs" compromise the ground system of the equipment. Several years ago, botanist Dr. Tarun Mal, a faculty member at Cleveland State University (Ohio, US), was found electrocuted in his lab, the victim of a power supply fault in his ungrounded, home-built plant-grow lights [55].
- When it is necessary to handle live electrical components, **use appropriate tools and personal protective equipment**. Tools with insulated handles, as well as electrically-insulative gloves and boots, are available, although insulated gloves and boots require periodic testing (on the order of once or twice/year) to ensure that they have not been compromised.
- **Avoid using extension cords.** There is an inherent potential for trip hazards, and damage to the cord (which can easily occur if the cord is tripped over, stepped on, run over by equipment, or fastened too tightly to the wall) can create hot spots that can result in fires. Extension cords should not be used as permanent wiring—have a wall outlet installed in the required location if the cord is needed for more than a few months.
- **Use power strips and surge suppression strips only when necessary.** Power strips and adapters multiply the number of available outlets, while surge suppressors absorb spikes in line voltage. (They are often combined into multi-outlet surge protectors.) Overloading the current-carrying capacity of the strip or suppressor can lead to a fire—always calculate the total load plugged into a strip and ensure that it

is no more than 80% of full rating. Use surge suppressors only for sensitive electronics. Connecting devices such as refrigerators or large motors can destroy the surge-protection capabilities due to large inrush currents when the device is switched on.
- **Always de-energize and lock out power to equipment while it is being serviced.** When you do not have continuous personal control over the power plug (or the equipment is hard-wired), obtain specialized training and follow written procedures when you will have access to potentially-energized components or when accidental startup of the equipment could injure you [56].
- **Never work alone with equipment that can store substantial amounts of electrical (or other) energy, or when servicing older equipment whose power source cannot be "locked out."** If there is no power disconnect that can be locked in the off position, someone who cannot see you may start up the equipment and electrocute you.
- **Never connect the neutral conductor to ground.** Consult a licensed electrician should this be necessary—it is not normally done on the building side of an electrical system, and special methods are necessary to prevent killing you or electricians working elsewhere in the building.
- **When grounding equipment at multiple points, be certain the ground is effective**. Earth conductivity can be quite low, and each "ground" may be at drastically different voltages, leading to effectively ungrounded equipment (due to so-called "ground loops"). This effect can also destroy your equipment or ruin your experimental results. Whenever possible, use the building ground system or a "single-point ground."
- **Consider the resonant frequency of any ground conductors when working with alternating current, especially in the radio-frequency range.** A 2 m ground wire is **not a conductor** for 144 MHz AC—it is an antenna with a substantial radiation resistance.
- If you or others in your lab will be working with energized components of electrical equipment, **you should receive cardiopulmonary resuscitation/automated electric defibrillator (CPR/AED) training.** This is normally available from local occupational health clinics, safety offices, and the Red Cross/Red Crescent.
- **Have and practice a plan for responding to electrical incidents in the laboratory.** You must not risk yourself attempting to pull a victim away from a power source, and you have only 2–3 minutes to begin CPR in order for it to be effective.

6.6.2 Magnetism

Electrical current generates a magnetic field. Some laboratory devices are designed to create very high magnetic fields: 1 T or higher is not unusual for nuclear magnetic resonance (NMR) spectrometers or magnetic resonance imaging (MRI) devices, and 45 T magnets exist in certain specialized facilities. Magnetic fields can have unusual biological effects, and magnetic field gradients create severe physical hazards due to their interaction with ferromagnetic materials.

6.6.2.1 Biological effects

Strong magnetic fields can cause immediate injury in the body, which is rich with calcium, magnesium, and other ions. Thermal and other motion of these ions (through the circulation of blood, for example) across a magnetic field leads to induced currents, and since the body is not a perfect conductor, localized heating results; if the magnetic field is strong enough, these currents can be damaging. Nerves may also be irritated or damaged due to similar inductive effects.

Sensitivity to electromagnetic fields is highly frequency-dependent. Due to disagreement among researchers (and government authorities) on the exact mechanisms responsible for this, particularly with weaker (mT-range) fields, exposure limits for electromagnetic radiation vary by several orders of magnitude. See [57] for further details regarding international electromagnetic field exposure standards.

6.6.2.2 Physical hazards of magnets

High magnetic fields (in excess of 3 mT or so) interact strongly with ferromagnetic materials such as steel. When brought within a few meters of a strong magnet (the distance increases with the field strength), these interactions are strong enough to pull in steel tools, equipment racks, etc., with sufficient kinetic energy to injure or kill researchers, damage the magnet (leading to quenching—see below), and cause shrapnel damage to nearby persons and equipment. Strong magnetic fields are also capable of erasing or damaging magnetic data storage media (hard drives, credit-card magnetic stripes, etc.).

Magnetic fields typically reach through unshielded walls, and it is possible for ferrous objects to be drawn **completely through the wall** if their attraction to the magnet is strong enough.

Superconducting magnets—commonly used for NMR and MRI—are cooled to superconducting temperatures using liquid helium and/or liquid nitrogen. Should an electrical or temperature-control failure allow the magnet to warm above the superconducting transition temperature, the magnet will suddenly become non-superconducting. Resistive heating by the electrical currents circulating in the magnet (and the subsequent collapse of the magnetic field) boils off a great volume of cryogen, potentially creating an oxygen-deficient atmosphere very rapidly (see section 6.4.3 above). This process is called **quenching**. Magnet damage from impact (e.g., a wrench drawn into the magnet) can also cause quenching. Rooms containing superconducting magnets should have fixed oxygen-deficiency detectors (which should be regularly tested and maintained), and occupants should be trained to evacuate immediately should the alarm sound.

6.7 General environmental hazards

In addition to having scientific and engineering experimentation, a laboratory is like any office or work environment and so has similar general hazards.

6.7.1 Trips, slips, and-falls

Falls are a significant hazard in the workplace. When we think of fall hazards, we normally think of workers falling off roofs, from ladders, from scaffolding, etc. These **fall from upper level** incidents are not the only sort of falls that occur, either at home or at work.

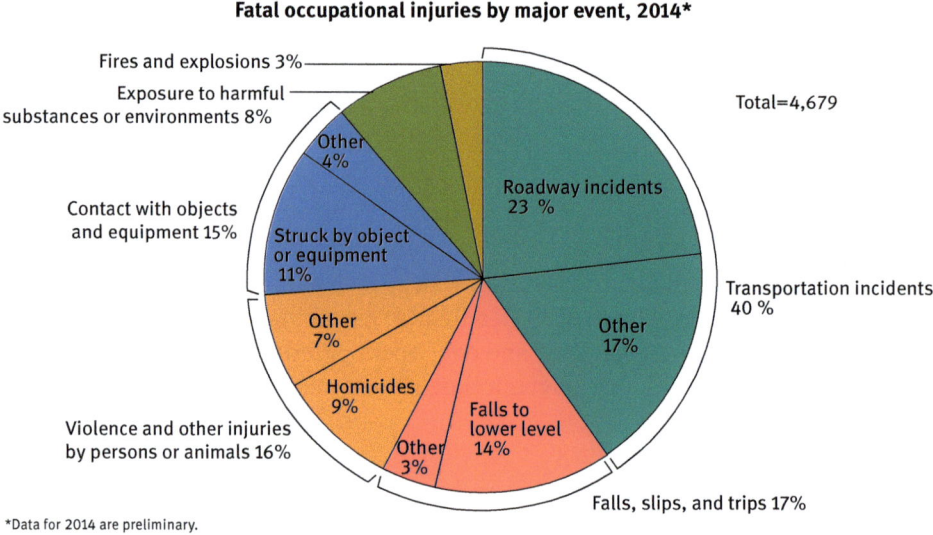

Figure 6.22: US Bureau of Labor Statistics Fatal Occupational Injury and Illness data, 2014. Figure reproduced from public domain BLS copy.

The US Department of Labor's Bureau of Labor Statistics tracks all workplace deaths in the United States; Figure 6.22 details fatalities in American workplaces for 2013. While violence and transportation incidents form the majority of incidents, falls are a substantial contributor—17% of all deaths at work. Note that "falls to lower level" is only 14% of the total, leaving 3% of workplace fatalities as **fall on same level**. An example of this type of incident sequence would be:
- Worker trips over filing cabinet drawer carelessly left open in a walkway;
- Worker strikes head on corner of desk while falling; and

– Worker dies of a subdural hematoma (blood collecting within and exerting pressure on the brain).

Given that there were approximately 4,700 deaths while at work in the US in 2013, this equates to **140 persons** who died tripping over trash bins, slipping on wet spots on the floor, etc.

Keep laboratories and supporting spaces clear of debris, cords and cables running across the floor, etc., and ensure that you do not leave drawers, trash bins, or other objects such as compressed gas cylinders in walkways.

6.7.2 Lighting

Although **lighting** is rarely the proximate cause of an accident, it is often a contributing cause. Poor lighting can make it difficult to see the extension cord strung across the walkway, or make it difficult for a researcher to tell the difference between a regulator designed for oxygen service and one designed for hydrocarbons. Professional associations and trade groups such as the Illuminating Engineering Society of North America (http://www.ies.org) publish guides to appropriate illumination levels and design for various work environments.

6.7.3 Noise

Much like lighting, **noise** can contribute to incidents by causing distraction and impeding communication between researchers. Noise can also be a hazard to hearing in and of itself. Most people think of proximity to a running jet engine or a rifle shot when they think of noise hazards, but lower volumes can cause just as much hearing damage when the exposure takes place over a longer time. Exposure limits for noise depend on frequency and time of exposure—consult an industrial hygienist, occupational health physician, or other health and safety professional for advice if noise is an issue in the lab.

6.7.4 Security hazards

Many research environments are relatively open to the public, particularly those located on university campuses. In such areas, security is also a safety concern. Thefts of purses, backpacks, and laptop computers do occur in laboratories, just as they occur in offices. Vandalism by outsiders can cause a great deal of damage to expensive, sensitive equipment as well.

Security issues raised by those working **inside** the laboratory area or building can also be problematical. In some laboratories, competition between students or other research workers can lead to sabotage of experiments—spitting in valuable cultures, spiking products with unwanted compounds, etc. Some such activities are merely "business interruption," that is, destruction of intangible property such as the results of an experiment, but some are threats to life and property.

Workplace violence is a risk in any workplace, including laboratories, regardless of location. Several incidents of attempted murder by poisoning with thallium salts (quite toxic) have occurred in chemistry laboratory facilities.

Keep laboratory doors closed and, if possible, locked, particularly when the lab is unmanned. Report suspicious occurrences to the proper authorities for investigation— usually the principal investigator in cases of minor sabotage and the local police in the event of significant property destruction or threats to personal safety.

6.8 Case study: Chemistry experiment

Emergent phenomena are those which arise from the action of groups, as opposed to phenomena which are due solely to individual items. Flocks of birds often present emergent phenomena, with the flock exhibiting behaviors not found in individual birds. An emergent phenomenon, or **emergence**, depends on interactions between simple rules followed by members of the group, whether they be birds or electrons. Not surprisingly, the **connectivity** between the group members determines the effectiveness of this cooperative behavior.

Many properties of materials are emergent phenomena, notably **superconductivity**. In a superconductor, electrons act as one in a manner evocative of the motion of a bird flock. Normally obstructions in the material (nuclei, impurities, dislocations in the lattice, etc.) introduce electrical resistance. Due to quantum effects, electrons pair up under certain conditions in so-called **Cooper pairs**, and small excitations such as those caused by obstructions are quantum-mechanically forbidden. Electrons thus flow without resistance: superconductivity. (For a more complete and correct explanation see [58].)

One class of materials potentially useful for the study of superconductivity, particularly at higher temperatures (i.e., higher than about 30 K) is **honeycomb iridates** [59]. These compounds promise access to an emergent phenomenon called **the quantum spin liquid state** critical to the understanding of high-temperature superconductivity. Synthesis of these materials proceeds by hydrothermal and other standard solid-state techniques. The different synthetic techniques can be used to modify the crystal lattice, adding and removing atoms, transforming the crystal symmetry, and changing bond lengths.

Though this materials research relies upon wet chemistry, the actual conduct of the syntheses and workup of the product involves several serious physical hazards. In order to set up a solid state reaction, one generally must use a high-pressure press to bring the ingredients into close contact. Powders ground by hand in a mortar and pestle are loaded into a pellet die. The die is placed into a press, and the press is hand-pumped to a high pressure to form a pellet from the ingredients. The resulting pellet is flame-sealed into a quartz tube using a methane-oxygen torch, often on a pressure-vacuum manifold used for work under inert atmospheres, and placed in a high-temperature (up to 1080 °C/1350 K/2000°F) furnace for hours or days. (See Figure 6.23.) Sometimes low temperatures are used for the reaction instead of using a furnace. The tube is removed, scored with a saw, and broken to extract the product pellet.

Figure 6.23: Furnace with blast shield.

Hazard controls for each of the above hazards are straightforward. The pellet press is equipped with a plastic guard to contain shrapnel or parts should the pellet die come apart under the high pressure. The lab uses only a small fraction of the pressure capability of the press, reducing the risk of disintegration. The flame work is performed in a dedicated area, and the torch is mounted on a stand for much of the process. Fuel and oxygen gas cylinders are restrained nearby, and cylinders are turned off when no torch work is going on.

Handling of hot samples from the furnace is done mostly with tongs to avoid contact. The presence of phosphorus or sulfur in the reaction raises the level of caution in the lab, as these mixtures sometimes explode during the reaction. In addition to conducting lower-temperature pre-reactions to reduce the risk, a blast shield is placed in front of the furnace and the tube is placed in a small metal shield inside to direct the force of any explosion away from the walkway in front of the furnace. (See Figure 6.23.)

Figure 6.24 shows a sample of the product, still in its quartz tube.

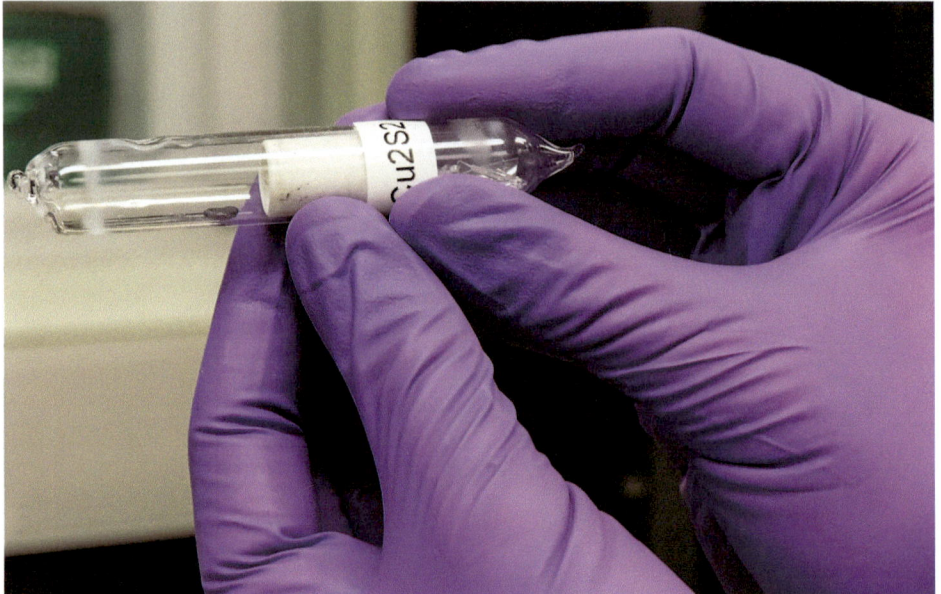

Figure 6.24: Synthesized material in quartz tube.

6.9 Exercises: Physical hazards

1. Match the physical hazard with an appropriate control measure.
   Hazard
   a. Asphyxiation by nitrogen gas.
   b. Electric shock from frayed electric cord.
   c. Cut from razor while forcing closed the sharps container.
   d. Hand/arm pulled into rotating vacuum pump shaft.
   e. Fire from hot torch set down near combustibles after use.
   f. Condensation of liquid air in open-top nitrogen Dewar flask.

g. Unidentified chemical in lab refrigerator.
h. Broken arm from tripping over equipment cord or cable laid across walkway.

Control
1. Clear work area of all combustibles before performing "hot work."
2. Inspect electrical cords periodically.
3. Do not "top off" Dewar flasks open to atmosphere.
4. Use inert gases in a well-ventilated area.
5. Keep motor & pump shafts guarded when energized.
6. Label and log all samples stored in lab refrigerators or freezers.
7. Keep all materials out of laboratory aisles.
8. Do not fill sharps containers more than 80% full.

2. Examine the following brief laboratory situations, identify physical hazards present, and give at least one practicable control to reduce risk from the hazards.
   A. A large 480 V electric motor fed by a frayed cable wired directly into the building power system is used to drive a wave generator in a mechanical engineering laboratory that studies dispersion of oil in salt water (such as occurred in the Gulf of Mexico/**Deepwater Horizon** oil spill in 2010).
   B. A chemical laboratory uses a pressurized metal reactor to hydrogenate sample compounds. The reactor, which is located in a chemical fume hood, is fed through plastic tubing by a cylinder of hydrogen clamped to the lab bench across the walkway.
   C. In a biomedical facility, a researcher uses a small machine shop containing a lathe, bandsaw, and drill press to construct experimental prosthetics.

7 Chemical hazards

Chemical hazards, as the name suggests, are hazards arising from the intrinsic properties of chemicals. At root, **chemical reactivity** accounts for all types of chemical hazard—although they are typically classified into subcategories, the difference between subcategory is the exact type and mechanism of reactivity.

In this chapter, we will discuss the four basic types of chemical hazard, primary methods for communication of chemical hazards, and protection against exposure to chemicals.

7.1 Routes of exposure to chemical hazards

In order for a chemical exposure to occur, the chemical must reach the victim. There are four main **routes of exposure** for chemicals:
- **Absorption:** A chemical splashed upon the skin may penetrate at various rates, some nearly instantaneously and some over very long periods of time;
- **Inhalation:** Compounds that are inhaled into the lungs are often easily absorbed through the linings of the organ, entering the bloodstream directly, allowing the compound to reach sensitive organs rapidly;
- **Injection:** Penetration of the skin via needles, syringes, contaminated broken glass, scalpels and the like also offers direct access to the bloodstream;
- **Ingestion:** While investigators do sometimes swallow chemicals in the lab (and in the 19th and early 20th century it was common to report the taste of a new compound in the literature), this sort of exposure also occurs when a chemical splash enters the mouth or nose.

7.2 Chemical properties contributing to hazard

Many chemical and physical properties of chemicals contribute to hazard. Two principal properties of chemicals determine much of the hazard presented by the compound: **reactivity** and **volatility**.

7.2.1 Reactivity

Chemical reactivity is the most obvious factor. **Reactivity** is used here in a very generalized way—chemical reactions figure into all of the principal classes of chemical hazard. A compound might:
- React with oxygen or other oxidizers, presenting a **flammability hazard;**

- React with the body to destroy or alter biochemical systems, causing a **toxicity hazard;**
- React with metals or with other electron donors or acceptors, creating a **corrosivity hazard;**
- React more generally with other compounds, or even with itself.

The term **reactivity hazard** is typically reserved for the last case, but flammability, toxicity, and corrosivity all result from chemical reactivity of one kind or another.

7.2.2 Volatility

A less obvious property that has a strong influence on hazard is the **volatility** of the compound. A more volatile chemical presents a greater hazard to users than a less volatile one, if all other factors are equal. As the vapor pressure of the material increases, the concentration in air increases. Inhalation exposures depend upon the vapor concentration—more vapor, more exposure.

Reactivity (in the form of toxic effects) and volatility are combined to provide a semiquantitative measure of chemical hazard called the **Vapor Hazard Ratio** (see page 129 below).

7.3 The chemical fume hood

One of the most important tools in the laboratory for preventing exposure to chemicals is the **chemical fume hood** (see Figure 7.1). The "hood," as it is usually called, exhausts air from the work area, preventing the researcher from breathing fumes, vapors, and noxious gases produced by the chemicals in use. The sash of the hood also provides some protection against splashes and minor explosions, although this should not be relied upon—if these are a serious possibility, a blast shield and face shield are recommended (as are other precautions, such as working on milligram-to-microgram scale).

A hood is a large metal cabinet (approximately 1–3 m; 3–9 ft wide) usually mounted on a lower cabinet or stand so as to provide a convenient waist-level working surface. See Figure 7.2 for a schematic drawing. A movable sash, usually tempered glass, is provided at the front of the cabinet, and recommended practice is to lower the sash as far as possible consistent with being able to work without hindrance. Air is drawn into the hood, by a fan atop the hood (an older design) or by a building hood exhaust system, and exhausted through the roof or at another convenient point where the exhaust air is not likely to re-enter the laboratory building or return to ground level. Various appurtenances may be provided inside the hood: cup sink; gas, water, or other fluid taps; electrical outlets, equipment support grids, etc.

88 — 7 Chemical hazards

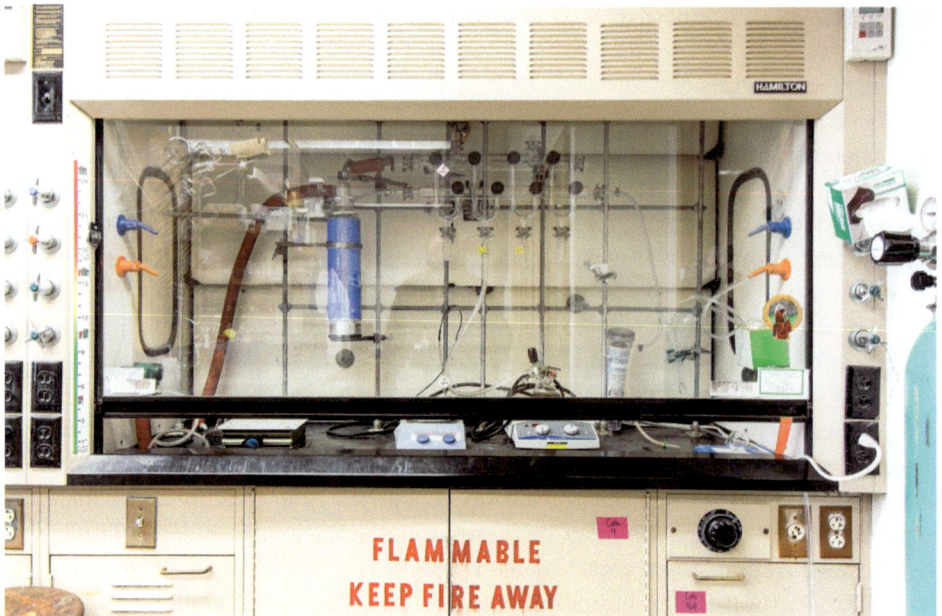

Figure 7.1: Chemical fume hood with Schlenk manifold.

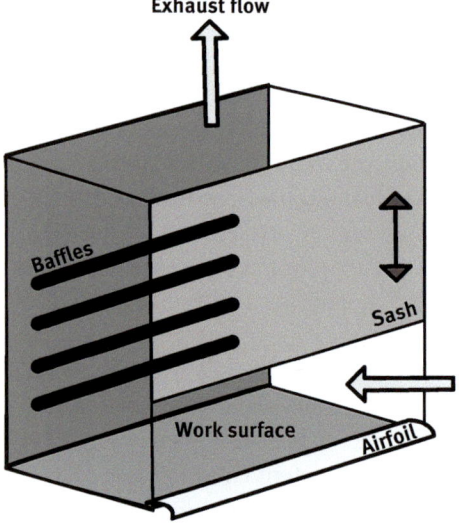

Figure 7.2: Chemical fume hood schematic.

> It is essential to understand that **a chemical fume hood operates best when it is empty and no one is standing in front of it.** Everything the researcher does, from placing equipment into the cabinet to performing work inside of it, degrades the effi-

ciency of the hood. It is **essential** to avoid disturbing the hood's airflow as much as possible to maintain the greatest protection.

The chemical fume hood is invaluable for reducing chemical exposure to inhaled substances in laboratories, but many factors can reduce its effectiveness:
- **Too high or too low an airflow through the hood**—either condition impairs the hood's ability to capture contaminants and prevent them leaving the cabinet. Inlet flow that is too low will permit materials to diffuse out against the flow, while too high a "face velocity" will cause excessive turbulence that draws vapors out. Newer hoods will often have an airflow meter that measures flow rate and sounds an alarm should it become too high or too low.
- **Working too close to the front of the hood**—as one works, one casts a wake, much like a speedboat in water. The turbulence of this wake is easily capable of pulling material out of the hood. Note that when contaminants leave the hood in this manner, they will be drawn directly into the area directly in front of the body from nose to chest (the so-called "breathing zone") and then pulled easily into the lungs. Work approximately 20 cm (6–8 in) from the front of the hood to allow your wake to dissipate.
- **Blocking the airfoil at the front of the hood or the baffles at the rear**. Both structures serve to distribute airflow through the hood and capture contaminants efficiently.
- **Crowding the hood with excessive equipment, chemicals, dirty lab ware, trash, etc**. Again, this disturbs the airflow and limits the ability of the hood to remove dangerous vapors.

A chemical fume hood is not proof against all incidents inside the hood. In 1993, a postdoctoral fellow at a major research university experienced a chemical explosion in his chemical fume hood. It blew a hole in the rear of the hood into the neighboring corridor and blew the glass sash into his face; opthamological surgeons removed about 50 pieces of glass from his eyes. Needless to say, the incident was a career-ending injury.

In another incident twenty years later, a graduate student started (in a chemical fume hood) a chemical reaction that exhibited an **induction period**, that is, it took some time before the reaction completely began. Being unaware of this property, the student continued dosing ingredients into the apparently quiescent reaction apparatus. He then raised the hood sash in order to adjust the equipment—and the reaction began vigorously. The resulting "burp" of boiling solvent forced the glass joint at the top of the reaction vessel up, spraying the researcher in the face with the reaction mixture, primarily methyl iodide. By chance, the experimenter's mouth was open at the time, and he received inhalation, absorption, and ingestion exposures from the facial splash. Fortunately, the exposures were too small to cause concern, but a call to the local Poison Center was necessary to determine the lack of danger.

7.4 General hazard classifications and precautions

Chemical hazards are organized into four basic types, mainly around the type of chemical reactivity that offers the hazard:
- **Flammables and oxidizers** participate in fires and explosions;
- **Corrosives** react strongly with skin and other exposed body tissues to cause large-scale destruction of tissue;
- **Reactives** react strongly with various other chemicals (or even themselves), releasing potentially large amounts of energy in the process;
- **Toxics** interact more subtly with body systems to interfere with proper operation. Several subclasses of toxic are often distinguished:
- **Carcinogens** (e.g., dimethylformamide) cause cancer, although it can be essential to know in what species, at what doses, and over what time period in order to evaluate the actual hazard;
- **Suspect carcinogens** (e.g., diesel fuel) are believed to cause cancer, but the scientific evidence is not sufficiently strong to declare them known carcinogens;
- **Mutagens** (e.g., ethidium bromide) are known to cause changes in genetic and/or epigenetic coding and/or expression;
- **Teratogens or reproductive hazards** cause birth defects, sterility, and other reproductive disruptions, sometimes to later generations.

7.4.1 Experimental protocols for chemical handling

Certain recommended practices apply to handling of **essentially all chemical materials**.
- **Know the properties of the materials you will handle.** Read the Safety Data Sheet for the chemical (see section 7.5.4 below), and if there is any mention of special hazards such as peroxide formation, spontaneous ignition, or reactivity with certain classes of other chemicals, do further research.
- **Wear appropriate PPE.** PPE must be compatible with the chemicals being used (recall the death of Karen Wetterhahn, who was unaware of latex's lack of resistance to the highly-hazardous dimethylmercury—see section 1.1.2 above.) Use PPE properly, for example, washing hands before removing eye protection so that the eyewear does not become contaminated. Chemical splash goggles are the only prudent choice for laboratories where chemicals that pose a threat to the eye are used. In fact, the American Chemical Society recommends splash goggles at all times in all chemistry laboratories [60].
- **Avoid ingestion of chemicals.** Food and drink, as well as food and drink containers, have no place in a laboratory (excepting a food science laboratory). It is not unknown for experimenters to accidentally ingest chemicals that have contaminated their coffee cups. Do not chew gum or apply lip balm, lipstick, or other cosmetics either, as this can trap contaminants from the laboratory air next to the skin.
- **Use the minimum amounts necessary for the experiment.** If your experiment requires a large amount of chemical (relative to the hazard), consider first how you might accomplish the experimental objective with less material.

- **Use blunt-tipped needles** where possible. Injection of experimental animals and puncture of rubber septa are the only typical reasons to use hypodermic needles in a research laboratory. Blunt needles will serve for drawing liquids from narrow containers and for other purposes. Once a rubber septum has been punctured once, the hole can be re-used with blunt needles.
- **Consider the energy to be released in the chemical use.** If you are washing parts with solvent, the primary hazards to be considered are absorption through your skin and inhalation into your lungs. If you are conducting an chemical reaction, runaway reaction and concomitant explosion is a possible consideration.
- **Dispose of waste chemicals properly.** In most areas, chemical waste must be labeled with the contents (full chemical name and amount) before it is surrendered to a licensed disposal contractor; keep track of what is placed in the bottle—"Organic Waste" is not an acceptable material to hand to waste disposal personnel. (At the author's institution, waste disposal is usually free—but unknowns result in a $500/€450 charge for analysis and characterization.)
- **Segregate waste properly.** Remember that chemicals are just as reactive in the waste bottle as in the reagent bottle or the reaction flask. Do not mix potentially-reactive materials in waste containers [61].
- **Transport chemicals properly.** Use damage-resistant secondary containment such as bottle carriers to move chemicals between laboratories (and longer distances in labs). **Never** carry a large bottle by the glass ring molded into the top—these rings can break spontaneously and without warning due to unannealed stresses in the glass. Carry larger bottles with two hands: one supporting the neck to stabilize the bottle, and the other supporting the bottom and carrying the weight.
- **Store chemicals properly.** Many chemicals are incompatible with other chemicals (acids and bases, for example, are mutually incompatible). Avoid storing incompatible chemicals in the same cabinet or on the same shelf. Several systems exist to classify chemicals for storage. Some of these systems have been created by manufacturers and appear on chemical bottle labels. The author favors the system that originated at Stanford University [62].
- **Be prepared for spills.** Have sufficient spill cleanup supplies **appropriate to the chemicals in use** on hand and know how to use them properly in the event of a spill. Do not use paper towels to soak up chemical spills. Special sorbents are available for oily or flammable materials, for acids and bases, and for special spills such as mercury. The author typically recommends keeping a large bag of sodium bicarbonate as extra spill control material for corrosives—the bicarbonate ion is **amphoteric** and can neutralize both acids and bases, and it is compatible with most corrosives. (Check the Safety Data Sheet for compatibility restrictions.)
- **Be ready in the event of fire.** Fires are uncomfortably common when handling chemicals. While you should work to avoid fires as much as possible (see section 7.4.2 below), also have a plan for fighting small fires and escaping larger ones. In many cases, the appropriate response to a fire in lab is to evacuate immediately, pull the building fire alarm, and call for professional firefighters. The smoke from laboratory fires tends to be even more toxic than that from typical building materials, and an amateur firefighter can easily be overcome.

7.4.2 Flammables and oxidizers

7.4.2.1 The fire triangle/tetrahedron

In order for a fire to occur, several conditions must be present. First, there must be a fuel source; obviously, fires do not burn if there is nothing to burn. Second, a source of oxygen or another compound that plays the same chemical role must be present—an

oxidizer. Third, There must be a source of ignition, whether a match, a hot surface, or even a spark of static electricity. Lastly, there must be some self-sustaining chemical reaction, begun by the ignition source, between the fuel and the oxidizer.

Fire educators represent these four conditions schematically as **the fire tetrahedron**. (See Figure 7.3.) Sometimes the need for a chemical reaction is omitted, and the diagram is called the **fire triangle**.

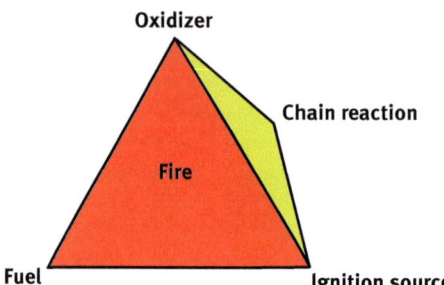

Figure 7.3: Fire tetrahedron.

Fire protection consists of ensuring that at least one of the four sides of the fire tetrahedron is not present. Without any individual side, the figure collapses and there is no fire. Consider fire prevention in an office:
- Removing large agglomerations of paper and other combustibles reduces the amount of available fuel;
- Forbidding smoking removes a possible ignition source.

It is frequently impossible to remove certain factors. In the office example above, removing oxygen to deny a fire an oxidizer is clearly impractical, and the laws of nature dictate that there is a possibility of a self-sustaining chemical reaction between paper and oxygen. If this is the case in a particular application, pay special attention to eliminating or reducing those portions of the fire tetrahedron that remain.

Incipient fires and fire extinguishers
Fire extinguishers exist in most laboratories, often due to fire code and other legal requirements. Whether researchers should be using them to extinguish fires, though, is an open question. There are several disadvantages to fire extinguisher use by non-firefighters:
- Generally, the use of an extinguisher is the first action taken by the person who discovers the fire, delaying sounding of the fire alarm or other evacuation system;
- Fire extinguishers have a limited charge, and training and experience is necessary to be able to accurately judge when a fire is too large and evacuation is necessary;
- Use of extinguishers is not simply a matter of aiming and squeezing the handle— poor technique can spread the fire instead of extinguishing it;

- Because the extinguisher user delays his or her evacuation in order to fight the fire, there is an increased chance that a sudden increase in the fire's size could cut off escape;
- All fires produce toxic smoke. In a laboratory situation, the toxicity can be much greater, as inherently-toxic materials are volatilized and added to the smoke. Inhalation exposures during laboratory fires are potentially fatal.

The author **does not recommend** use of fire extinguishers in laboratories except for the most trivial (e.g., trash can burning) fires or in extreme cases such as in response to blocked escape routes. Firefighters in the United States have a saying: "the lifespan of an amateur firefighter is one fire," and statistics bear this stance out: Of those who die in residential structure fires, a large proportion of them were attempting to fight the fire at the time, and some had even re-entered the structure in order to do so [63].

7.4.2.2 Flammable vs. combustible
The words **flammable** and **combustible** are used interchangeably by the public, but there is a difference:
- **Combustible** means merely that a substance can be burned. Cardboard, wood, cotton, heavy crude oil—these are all combustible;
- **For liquids, flammable** means that the **flash point**—the temperature at which the vapor immediately above the surface of the material is ignitable—is at or below the current temperature of the material.
- **Flammable gases** are ignitable in certain mixtures with air. Usually we express the range of flammability for gases in terms of the **lower flammable limit** (the lowest concentration that will burn—below the LFL, the combustion releases too little energy to be self-sustaining) and the **upper flammable limit** (the highest concentration that will burn—above the UFL, there is too little oxygen present to continue the combustion reaction). The LFL and UFL are also known as the **lower and upper explosive limits** (LEL and UEL) respectively.
- **Flammable solids** burn above a certain rate under laboratory conditions, or can be self-igniting. Some flammable solids provide their own oxidizer, such as ammonium nitrate fertilizer.

Various fire regulations set numerical limits on the degree of flammability (e.g. LFL < 14% to be qualified as a flammable gas in the US for transportation purposes) or divide materials into flammability classes according to the degree of hazard or ease of ignition.

7.4.2.3 Ignition sources in the lab
Several things can provide the ignition source for a laboratory fire:
- **Electricity**, whether from equipment or a static discharge;
- **Deliberate ignition** with a flame source such as a match or flint-and-steel;

- **Hot surfaces or equipment**, such as a sample just removed from a furnace, or the tip of an acetylene torch used to seal glass ampoules; and
- **Exothermic chemical reactions.**

Electrical ignition
Electricity, whether from apparatus or from a static discharge, is a common source of ignition. Faulty electrical equipment, a deliberately-created spark, or even the act of plugging a piece of equipment into a wall socket can ignite flammable materials. Insulating liquids such as gasoline can also build up static charges when poured in quantity; upon discharge, the static can easily ignite the liquid and cause a fire or explosion.

When pouring combustible/flammable materials from a metal or other conductive container, always **ground** the container to a suitable electrical ground and **bond** the container to the receptacle receiving the liquid (connect it with a low-impedance connection). This places all equipment at the ground potential and removes the risk of a static-ignited fire.

Deliberate ignition
Laboratories use supplies and equipment intended to ignite fires: matches, flints, lighters, etc. Lighting a lab burner is a particularly common operation.

Clear all benches, chemical fume hoods, and biological safety cabinets of flammable materials before lighting Bunsen or Fischer burners or torches. Also check the lab for open containers of flammables or other materials that could ignite—there are anecdotal reports of open bottles of diethyl ether (an extremely flammable substance) igniting more than 5 m (20 ft) away from a lit burner. Flammable vapors can travel quite long distances.

Ignition by hot surfaces/equipment
Furnaces, hot plates, samples removed from furnaces or kilns, and other hot equipment are common sources of lab fires.

In a recent incident at the author's institution, a researcher flame-sealed an ampoule of chemicals before removing it from a vacuum manifold. She set the hot torch down on the work surface of the chemical fume hood and left to stow the ampoule. When she returned, the torch had ignited an agglomeration of wipes and rubber septa left in the hood, and the fire was about to spread to a solvent still located in the hood.

Never use open flames or hot surfaces above approximately 200°C (400°F) in an area containing flammable materials, even if the area is well-ventilated (like a fume hood).

Ignition by exothermic chemical reactions
Some chemical reactions release large amounts of heat—combustion reactions being a clear example. Synthesis of "self heat-sustained" (SHS) ceramic materials is another example. The exterior of containers for mixtures undergoing exothermic reactions can become hot enough to serve as an ignition source. Do not conduct such reactions near flammable or combustible material. (It should be noted that some of such reactions **involve** flammable materials as reactant or solvent. Take care to conduct such experiments with due regard for the possibility of a larger-scale ignition or fire due to a faster reaction rate than anticipated or equipment breakage.)

7.4.2.4 Fuel sources
Many types of materials can serve as fuel sources. Common flammable or combustible laboratory materials include:
- **Flammable chemicals and solvents**, whether used for "chemical" purposes or for operations such as parts-cleaning;
- **Laboratory wipes, absorbent pads**, etc.;
- **The contents of laboratory trash containers**;
- **Metal or wood dust agglomerations** in shop locations;
- **Combustible metal** powders (some of these are capable of self-ignition if fine enough and exposed to the air);
- **Flammable gases**, including natural gas supplied at the bench or hood.

7.4.2.5 Storage of flammable chemicals
Follow the practices below to safely store flammable materials in the laboratory:
- **Store all flammable materials in flammables storage cabinets** approved by the local fire marshal or other authority. Small amounts (a few containers < 4 L/1 gal each) can sometimes be stored on the laboratory shelf if local laws permit. Any large container other than an approved safety can belongs in a flammables cabinet (e.g., a plastic 20 L/5 gal mini-barrel of ethanol, such as might be kept on hand in a biological laboratory).
- **Do not store flammables in chemical fume hoods, on the bench, or in sinks.**
- **Remove materials from the flammables cabinet only when about to use them**. Replace them when you have withdrawn the amount you intend to use.
- **Flammable materials that are also corrosive (e.g., acetic acid) should be stored in a flammables cabinet**, not in a corrosives-storage area. At a minimum, isolate the corrosive in a separate secondary container to minimize damage to flammables cabinets, which are generally metal, and seal containers well. If the corrosive is volatile, wrap the lid of the bottle with paraffin wrap or other sealant to reduce corrosive vapor release.

- **Use flammable materials only in well-ventilated areas.** In the typical laboratory, this means only in the chemical fume hood. Bench use of flammables is inadvisable.
- When using highly-volatile flammable materials such as diethyl ether, first survey the lab and verify the absence of ignition sources, and notify other researchers that the material is in use. **Volatile flammables can ignite over quite long distances.**
- Experiments or operations involving volatile flammables or highly-flammable gases can sometimes require **special ventilation and electrical apparatus.** Check your local fire code or consult with a fire-safety specialist if you suspect this may be the case. Generally, this happens in situations where there may be a buildup of flammable gases or vapors.

7.4.2.6 Oxidizers

As mentioned in section 7.4.2.1 above, an **oxidizer** is a material that contributes to combustion by playing the role of oxygen in the reaction. Technically, an oxidizer acts to accept electrons from the fuel; the overall transfer typically releases large amounts of energy.

Common oxidizers

Laboratory oxidizers may be solid, liquid, or gaseous:
- **Oxidizing gases** include oxygen, nitric oxide (and other nitrogen oxides), and halogens (i.e., chlorine, fluorine);
- **Oxidizing liquids** include bromine, nitric or perchloric acid, and hypochlorite or chromate solutions, or liquified oxidizers such as liquid oxygen;
- **Oxidizing solids** include iodine (a mild oxidizer), potassium permanganate, and various organic and inorganic peroxides (e.g., benzoyl peroxide).
- **High-pressure or liquid cryogenic air** can even act as a strong oxidizer.

Precautions for handling oxygen and other oxidizers

Observe the following practices when handling oxidizing chemicals:
- First and foremost, **avoid any possibility of contact between strong oxidizers and flammable or combustible materials,** unless doing so is part of the experiment and done under strictly controlled conditions by experienced personnel. Nitric and perchloric acids, as well as liquid oxygen, accelerate combustion drastically (because large amounts of oxygen are available very close to the site of combustion), and what would be a small laboratory fire can easily turn into an explosion. (Industrial spills of liquid oxygen from transport trucks have been known to cause the pavement upon which the "LOX" was spilled to explode.)

- **Many oxidizing chemicals can decompose rapidly upon heating.** Always research the properties of the specific material with which you are working. Store such compounds according to the manufacturer's recommendations, and hold them no longer than necessary (and certainly no longer than any expiration date given on the container).
- **Some oxidizing solutions emit gas upon standing because of slow decomposition.** "Piranha etch," a mixture of concentrated sulfuric acid and concentrated hydrogen peroxide used for high-grade cleaning of glassware and in electronics fabrication, is a notorious example. Allow such solutions to stand until gas evolution ceases. Once emissions have decreased to a low level, store the solution in waste bottles with vented caps.
- **Store oxidizers as far away from flammable materials as possible.** This prevents fires and explosions in the case of leaks—and note some leak sources, like a light fixture falling from the ceiling, could break both oxidizer and flammable containers simultaneously. Local codes often dictate separation distances, particularly for compressed gas cylinders of flammables and oxidizers. At the very least, store flammables and oxidizers at opposite ends of the laboratory.
- **Never** store oxidizing waste in the same waste container as combustible materials such as organics; explosions may result. It is essential to remember that just because a compound is designated as "waste" **does not make it magically lose its reactivity**. Mixing oxidizers and organics in a waste bottle is as dangerous (or more) than mixing them in a beaker.

At a university near the author's institution, two undergraduates in a beginning organic chemistry laboratory were left with a moderate amount of mixed sulfuric and nitric acids. They disposed of this material by dumping it in a bottle marked "Waste" provided at the front of the lab. Unfortunately, the waste bottle contained mixed organics, and it immediately exploded. Both students were taken to the hospital with minor wounds, and two members of the university staff required decontamination by local firefighters. Depending on the severity of the exposure and the hazard of the chemicals involved, "decontaminated" can mean anything from "wiped down with a wet-wipe" to "stripped and hosed down with a fire hose while being scrubbed with detergent." The latter is not an experience most people would covet.

Cleaning glassware with oxidizing chemicals
Some laboratories generate glassware with extremely difficult-to-clean encrustations or stains, or they require glassware to be completely free of metal contamination before use. Such labs make use of strong oxidizing solutions such as concentrated nitric acid (used to desorb metals from glass surfaces) to chromic acid, Piranha etch,

(or a substitute such as NoChroMix®—Godax Labs), and often one or more tubs of oxidizer is kept in the lab for immersing glassware.

Such operations should be carried out with extreme care. The presence of so much oxidizer poses a major hazard to personnel, and careful thought should be given to several factors.
- **Locate baths close to sinks or water baths used for rinsing, and as near as possible to the safety shower and eyewash.** Never place a bath on the floor where it may be tripped over.
- **Check materials compatibility tables, preferably from the manufacturer, before selecting containers.** Many plastics resist oxidizers to only a limited extent.
- **Use secondary containment trays to contain leaks and drips.** Note that a secondary containment tray should be at least the capacity of the bath container, lest a leak overflow the containment.
- **Use extreme care in immersing glassware in such baths. Never** plunge or drop items into the acid bath—lower them slowly. Splashes can cause very severe burns. Appropriate protective equipment (including impervious apron, chemical splash goggles, suitable gloves, and a face shield with chin guard) are recommended—strongly. Use plastic tongs to insert and remove glassware to be cleaned, not gloved hands.
- **Plan carefully for the procedure to be used for changing an oxidizer bath.** Filling and emptying the bath involves a large amount of hazardous material, with a concomitant increase in risk.
- **Always completely clean glassware before using an oxidizer bath.** Any significant organic material left on the surface, such as encrusted deposits, must be removed because it will react strongly with the bath, causing a splash or possibly an explosion. Cleaning with oxidizer baths is the last step, not the first.

Avoid using such baths unless absolutely necessary. Contrary to common misconception, it is often not necessary to maintain separate baths for each metal of concern to avoid cross-contamination—replacing the bath at reasonable intervals will generally prevent re-deposition. This is dependent upon the nature of the work and the purity required: analysis at part-per-trillion levels may very well require multiple rinses with fresh ultra pure nitric acid [64].

Base baths, particularly potassium hydroxide in ethanol, are sometimes also used for glassware final cleaning. The precautions above also apply to such situations.

7.4.2.7 Pyrophoric materials
Some materials are capable of self-ignition in air, that is, the temperature needed to ignite the compound is at or below room temperature. These compounds are called

pyrophoric materials and include such chemicals as **tert**-butyllithium, silane, and white phosphorus. Each of these compounds have been involved in deadly laboratory accidents.

Because of the self-ignition properties, pyrophoric materials must be handled with extreme care—researchers have died from exposure to centiliter quantities of pyrophorics (see section 1.1.6 above). Special techniques are required for using pyrophoric chemicals, and in the lab they are typically handled under air-free and moisture-free conditions (e.g., via a Schlenk apparatus—a glass manifold provided with both inert gas and vacuum for blanketing and transfer of air- and water-sensitive materials).

See [19] for further detail on pyrophorics handling, obtain appropriate training from an experienced person, and make dry runs before conducting any experiment involving pyrophorics.

7.4.2.8 Fire-retardant clothing

Clothing that **resists** fire is available for laboratory use. While certain materials used in industrial applications and by firefighters (including, formerly, asbestos fiber) are fireproof, materials generally used for laboratory fire protection are combustible but **self-extinguishing**. This means that they burn when a flame is applied, but when the flame (or other ignition source, e.g., an electric arc) is removed, they cease burning. This can be helpful if, for example, you are splashed with a small amount of burning solvent during an incident.

Researchers handling large (>1 L) amounts of flammables, who are exposed to electric arcs, or who handle **any** amount of pyrophoric materials should use fire-retardant laboratory coats or aprons. In larger engineering laboratories, fire-retardant clothing such as uni-suits or work shirt-and-pants combinations may be appropriate.

Fire-retardancy may be inherent to the fiber, such as with Nomex IIIA® (DuPont), or may be an applied fiber treatment, such as Proban® (Bruck), which is a cotton, cotton-nylon, or cotton-polyester blend treated to retard ignition. Treated fabrics tend to be much more sensitive to the method of washing; ammoniacal treatments, for example, may be destroyed by washing with chlorine bleach. Researchers should always arrange for commercial or industrial washing of laboratory wear by a laundry familiar with the fabric instead of washing it by themselves. (In addition to destroying the flame-retardancy, self-washing will likely contaminate the clothes washer with whatever was on the lab coat or possibly cause a fire. Seriously-contaminated clothing should be disposed instead of washed.)

Fire-retardant garments are more commonly rated for electrical arc flash than for actual resistance to flammable materials fires. If possible, select garments meeting NFPA 2112, a standard for clothing designed for use with flammables [65].

7.4.3 Corrosives

7.4.3.1 Mode of action

Acids and bases both act upon tissue through acid- or base-catalyzed hydrolysis of amides (including proteins) and esters (such as lipids). The resulting damage causes necrosis of living tissue, color changes, etc. Obviously, the effects are concentration-dependent—dilute acids act more slowly and less dangerously than concentrated ones. Some types of corrosive materials have special hazards:
- Very concentrated strong acids, such as concentrated sulfuric acid, can act as dehydrating agents, removing water from carbohydrates and other molecules, causing large-scale tissue destruction;
- Oxidizing acids (nitric, perchloric, concentrated sulfuric) destroy tissue via reduction-oxidation reactions, in addition to acid-base reactions, multiplying the hazard. Piranha etch (usually a 50% mixture of concentrated sulfuric acid and concentrated 30% hydrogen peroxide) is another type of "oxidizing acid."

  Never attempt to make Piranha etch using greater than 30% hydrogen peroxide. Concentrations of H_2O_2 over about 50% are inherently unstable and can decompose rapidly. Note also that Piranha etch should be made by adding the **hydrogen peroxide** to the **acid,** with gentle stirring. Adding acid to peroxide risks local concentrations over the 50% limit and may inadvertently cause the solution to erupt.
- Hydrofluoric acid (HF) is, chemically, a weak acid; most weak acids and bases do less damage to tissue than strong ones simply because the concentration of hydrogen or hydroxide ions is lower in the solution. Nevertheless, HF penetrates the skin quite readily, being carried into tissue and thence to the bloodstream. The fluoride ion forms insoluble compounds with divalent cations such as calcium and magnesium, effectively removing those ions from the tissue; if the fluoride reaches the bloodstream, this can occur systemically. This causes large-scale disruption in the functioning of cells, particularly nerve cells. Victims of HF poisoning frequently die hours to days later from cardiac failure, caused by cardiac nerve damage. Deaths are reported to have occurred after HF exposures approximately 5 cm (2 in) in diameter.

  HF is also a topical anesthetic, and exposures often go unrecognized for hours. HF should not be used without special training on the hazards, appropriate PPE, and on first aid for HF exposure. Safety Data Sheets and medical references on emergency treatment [66] sometimes list calcium gluconate gel as a suitable antidote to HF poisoning **if applied immediately**. It is inadvisable to use HF without an amount of calcium gluconate gel appropriate to the potential exposure on-hand (and unexpired—the shelf life of calcium gluconate can be relatively short). Facilities with a high risk for HF exposure should also ensure that the local Emergency Department is familiar with treatment and medical management of HF poisoning.

HF reacts easily with glass, and HF or HF-containing waste should not be stored in glass containers. Note that any acidulated fluoride-containing solution or waste, as well as solutions of materials such as ammonium bifluoride, forms HF in-situ, and also should not be stored in glass.

Gloving and chemical protective clothing

Since the body part most likely to be exposed to chemicals is the hands, **hand protection** using gloves is the most important personal-protective approach to chemical safety. **Always choose gloves that are appropriate for the chemical in use and the way it is used.** Consider the following situations:

– Wearing disposable gloves for handling large amounts of acid is dangerous. Use thicker **reusable gloves**, preferably with long, turned-up cuffs, when handling larger quantities. (The word **larger** in this context means "large relative to the hazard of the chemical.)
– Using nitrile gloves when handling acetone is also hazardous. Acetone penetrates nitrile easily [67] and metabolizes to carbon monoxide in the body. Headaches, flu-like symptoms, and more serious poisoning can result. Use glove materials that are resistant to the chemicals you are using.

Many Safety Data Sheets (SDSs) (see section 7.5.4 below) simply instruct the user to "wear appropriate gloves," an infuriatingly vague remark in most cases. Nevertheless, the chemical manufacturer, who provides the SDS, is **not** in the best position to recommend gloves for a particular application. Always ask the following questions when choosing gloves:

1. **What glove materials resist the chemicals in use?** Many common laboratory solvents, such as acetone and isopropyl alcohol, penetrate the common disposable nitrile laboratory glove. **Do not use disposable nitrile or latex gloves as a "default" protection in the laboratory without forethought.**
2. **How long must the glove withstand the chemical?** Depending on the particular operations you will conduct, some exposures may be more evident than others. If the time the compound takes to penetrate the glove is significantly longer than the time it will take you to detect an exposure, it may be appropriate to use those gloves, provided that they are washed and/or removed **immediately** after use. Inspect gloves before use, as a glove with pinholes will resist chemicals not at all.
3. **How long will you be using the chemical?** Typically, the penetration time for a chemical through glove material is directly proportional to the thickness of the glove. Do not wear disposable gloves for tasks involving long-term contact with the chemical (e.g., dipping glassware in a chemical bath, greasing parts.
4. **Are there any physical hazards that must be protected against?** Different gloves are differently vulnerable to damage—some may resist abrasion better than others, and others may be more cut- or puncture-resistant.
5. **Does the task to be performed require high dexterity?** If so, use thinner disposable gloves, **provided it does not create a higher risk of exposure**; if it does, redesign the task so that the lower dexterity of thick gloves is not an issue.

Information on chemical resistance of gloves to common laboratory chemicals is typically available from the manufacturer [67, 68], either in the form of literature or online web applications. In certain cases, outside agencies such as government occupational health & safety authorities may also provide advice. No glove can be tested for all chemicals; if you cannot locate chemical resistance information for a glove, consult a chemical safety professional, chemist, or similar qualified person, or simply choose another glove material.

Typical laboratory tasks require different gloves. Some rules-of-thumb:
- **Working with large quantities of chemicals (such as acid-bath dips or solvent parts washing) should be done with thick reusable gloves.** Turn the cuffs up at the ends so that any liquid that runs up the side of the glove is caught rather than channeled into the glove;
- **If there is a high probability of direct contact between glove and chemical, use thick reusable gloves.** Splashes to the gloves are likely to be unnoticed for a period of time, and the longer penetration time of thick gloves may be all that prevents an exposure;
- **If the compound in use is carcinogenic, a reproductive toxin, highly-toxic, or of unknown toxicity and likely to be extremely harmful or lethal (e.g., organomercury compounds), wear two pair of gloves:** an outer pair that is likely to resist the chemical, and an inner pair made of a multilayer polymer film (e.g., SilverShield®—Honeywell Safety Products; note that these products come in several different varieties, so choose the most protective for the application);
- **If the chemical is a common laboratory chemical for which data is available, and does not meet the criteria above, wear disposable gloves known to resist that chemical.** For low-hazard chemicals used in small amounts, used in ways that will not permit long-term contact between glove and chemical (e.g., where the gloves will not be dipped in the compound or splashes not noticed for a long time), general use of disposable nitrile gloves may be appropriate.

Note that these rules-of-thumb are generalized and may not apply to all situations. If you have any question about what glove to choose for a particular task, consult with a safety professional or with the manufacturer of the proposed gloves.

Eye protection for chemical hazards

As mentioned earlier (see page 52 above), safety eyewear designed solely to protect from impact, such as safety glasses or impact goggles, **is not appropriate for use with hazardous chemicals.** The only time such eyewear should be used for chemical applications is if the compound in use is a) **not** an eye hazard (corrosive, toxic, reactive, irritant) and b) is used in very small quantities. Safety glasses with side shields may, for example, be suitable for use in microbiology laboratories where mild buffer solutions and media are being handled. Note, though, that some biological buffer solutions contain toxic components or are corrosive, and media may be hot; in neither situation will safety glasses protect your eyes from a splash.

The standard eye protection for chemical hazards is the **chemical splash goggle**. (See Figure 7.4.) These goggles are designed with indirect ventilation paths that do not permit splashes of liquid to penetrate easily, instead of ventilation holes to permit easy air circulation. Goggles with no ventilation at all are sometimes used for applications involving exposure to hazardous gases and vapors; they are not preferred for other applications because they fog easily. Just as with impact goggles, models are available to fit over most corrective eyewear, and prescription inserts can also be found.

When choosing chemical splash goggles, look for several important features:
- Choose goggles that **fit close to the face, with a good seal between forehead/cheek and goggle**. Human head and facial shapes vary, and highly-curved models may not fit a person with a flatter face (and vice versa).
- Choose goggles that **have toric or cylindrical lenses**. The curve of the lenses reduces visual distortion at the sides of the viewing range.

– Choose goggles with **effective anti-fog treatments**. Many less-expensive goggles fog easily, while more "deluxe" models include scratch resistant, anti-fog lenses that make working in humid environments or under physically-demanding conditions much easier.

Figure 7.4: Chemical splash goggles (right) versus impact goggles.

Inexpensive chemical splash goggles can be found at home centers and hardware stores for a few dollars or euros; "expensive" goggles cost between ten and twenty dollars or euros (in small quantity). Buy the expensive goggles—inexpensive goggles are typically of low quality and are annoying to use—so they will not be used (and the money spent to purchase them will be wasted). In the author's experience, researchers are much more likely to wear a high-quality pair of goggles than a lower-quality one.

When using goggles, consider these hints:
- **Never** handle goggles with contaminated hands. Contamination will transfer to the goggle, and the next time the goggles are handled with bare hands, the hands themselves will become contaminated. Accidentally wiping one's eyes with hands carrying acid residue is not a pleasant experience.
- **Always** fit goggles to the face and then pull the strap over your head. This provides a good seal between face and eyewear.
- When removing goggles, **remove gloves, wash your hands, then stretch the strap backwards and slowly pull it over your head—then remove the goggles with both hands.** This technique helps prevent tangling the hair in the strap, and will also prevent corrective eyewear from being pulled off by the goggle, with a little practice.
- If your "anti-fog" goggles fog frequently, washing them and allowing them to air-dry will often fix the problem. If the anti-fog is a treatment applied with a spray or wipe instead of built-in to the lens, you will need to reapply.

7.4.3.2 Storage of corrosive materials

Storing corrosive material is relatively easy if a few simple rules are adhered to:
- **Do not** store corrosive materials above eye level. This makes it much more difficult to safely remove the container from the shelf, and increases the chance of a large splash to the face and upper body.
- **Hold corrosive containers with both hands, not by the neck of the bottle or the molded-in ring.** This reduces the chance you will drop the bottle (or the ring

will crack off). Transport containers between laboratories in secondary containment (a tray on a cart or in a plastic or rubber container tote).
- **Store acids separate from bases.** Some accidents in lab can break multiple bottles, and mixing of spilled acid and base can be unpleasantly spectacular. If only one corrosives cabinet is available, storing acids and bases in the same cabinet may be acceptable provided they are stored in separate secondary containment trays (this depends on local regulations).

  NEVER store oxidizing acids in the same secondary containment with **any other material**! Nitric, perchloric, concentrated sulfuric, and other oxidizing acids should be stored **individually** because of their strong reactions with other chemicals—reactions that often produce explosive by-products.
- **Use approved corrosive storage cabinets.** Do not store corrosives in flammables cabinets (except as noted below), on shelves or in standard reagent or tool cabinets—they are not designed to resist damage from corrosive vapors and may eventually collapse.
- **Store flammable corrosive materials in a flammables storage cabinet, not a corrosives cabinet.** The risk from fire generally exceeds the risk from corrosivity. Seal corrosive flammable bottles well to prevent corrosive vapors from being emitted—wrapping the outside of the joint between cap and bottle with paraffin film may be helpful.
- **Use secondary containment for all stored corrosives**. A secondary container should be large enough to contain the entire volume of the stored primary container, in case it leaks; low trays are not sufficient. For single bottles, a plastic beaker often works admirably for this purpose, although in the case of perchloric acid, a large glass beaker may be more prudent.

7.4.4 Toxics

7.4.4.1 Types of "toxicity"

Toxic (n.): poisonous
—Oxford English Dictionary (online version)

The word **toxic** has a simple definition that covers a wide variety of different compounds and mechanisms. For our purposes, **toxic** chemicals exhibit a type of chemical reactivity that interferes in some way with the metabolic or other biochemical systems of the body.

Acute versus chronic toxicity
An **acute** toxin causes immediate injury, that is, injury during or shortly after a single exposure of sufficient quantity. **Strychnine** is a strong acute toxin—around 200 mg is all that is necessary to kill a 70 kg (150 lb) human.

A **chronic** toxin causes damage over long periods of exposure. **Ethanol** (ethyl alcohol) causes damage to the liver over long exposure, ending in cirrhosis or hepatitis if drinking is heavy enough.

Exposure to **sensitizers** such as toluene di-isocyanate (TDI) may cause no visible effect upon initial exposure. Depending on dose size and frequency, later exposures may cause allergic-type reactions ranging from hives to shortness of breath to anaphylactic shock.

Benzene exhibits acute and chronic toxicity, depending on the dose. One of its more serious chronic effects is cancer (leukemia); a September 1948 American Petroleum Institute toxicological report stated, "...the only absolutely safe concentration for benzene is zero" [69]. Chemicals that cause cancer are called **carcinogens**. Cancer is usually a chronic effect, although some compounds, such as **ethylcarbamate** (urethane), are well-known to cause cancer after a single sufficiently-large exposure [70].

Cancer is the result of genetic and epigenetic changes induced by reactions with the carcinogen or one of its metabolites (breakdown products produced by metabolism in the liver or other organ), but cancer is not the only possible result of such changes.

Mutagens such as ethidium bromide, a commonly-used DNA stain for molecular biology, interact with DNA in ways that can cause mutations. Though mutations often lead to cancer, carcinogenicity generally requires multiple mutations at specific locations in the genome. Ethidium bromide is not currently believed to be carcinogenic, and it is even used as a veterinary drug in cattle (to treat trypanosomiasis, a parasitic disease). The mutagenicity of ethidium bromide is a matter of some dispute: it registers as mutagenic in the standard "Ames test," a cellular-level test for mutation-inducing ability, but animal studies do not confirm this.

Some chemicals cause mutations in germ cells (sperm and ova) that can be carried across into following generations, or can interfere with the proper development of a fetus. The latter reproductive hazards are called **teratogens**—they cause congenital defects such as multiple or no fingers (or hands), missing digestive systems, etc. (The word is roughly and unkindly derived from the Latin for "creator of monsters.") **Thalidomide** was used as a medication in the 1960's to treat morning sickness in pregnant women; despite intense political pressure from the manufacturer, a physician at the US Food and Drug Administration, Dr. Frances Kelsey, refused approval for the drug to be marketed in the United States, citing concerns about possible teratogenic effects. The drug was approved in several European countries and was soon found to cause a variety of birth defects; up to 10,000 children were affected before it was banned [71]. Ironically, thalidomide was later found to be suitable for cancer treatment, and it is used for that purpose today—but not in pregnant women.

7.4.4.2 Interpreting toxicity information

Toxicity is measured on many scales and in many ways, creating a very confused situation for the practicing scientist or engineer.

Toxic threshold levels

The toxicity of a chemical raises the natural question, "how much of this chemical is too much?" In other words, "how much of this chemical is required to observe toxic effects?" Because there are many types of toxicity (e.g., some toxins may target the liver, while others may cause cancer), many ways to be exposed (e.g, inhaled, ingested, etc.), and many different types of exposure (e.g., chronic occupational exposures vs. acute exposures to the public during emergency situations), there are many different "toxic levels" used. Those of most interest for laboratory researchers include:

- The **LD_{50}**. This is the dose at which 50% of a tested population (of animals, usually) administered that dose will die. ("LD" stands for "lethal dose.") While it can be useful in distinguishing one chemical from another—given the choice, work with the compound having the higher LD_{50}—the fact that many will die from that dose makes the LD_{50} non-useful as a target for controlling exposure. LD_{50} values, which are given in mg/kg body weight) apply to ingested, injected, or absorbed doses; the corresponding value for inhalation exposure is the LC_{50} (as parts per million by volume).
- The **NOEL**. This is the minimum dose for which toxic effects have been observed. (NOEL="no observed effect level.") While informative, it is **not a guarantee** that there will be no toxic effects observed below that level; the effects may simply be too rare to have been observed at a statistically-significant level in the tests conducted to date.
- The **LD_{LO}**. This is the lowest level at which fatality was observed. Like the NOEL, it does not represent a "maximum safe exposure."

The best use of toxicological values such as the LD_{50} to the practicing researcher is as a qualitative, comparative measure between chemicals. If a new compound with which you will be working has a similar LD_{50} to table salt, minimal precautions are probably required; if the LD_{50} is similar to strychnine (about 30 mg/kg body weight), more serious protective measures would be prudent.

Limitations on use of toxicity numbers

Toxicity values such as the LD_{50} should **never** be taken "at face value." The toxicity of a chemical depends on many factors:

- **Species.** Most LD_{50} studies are performed on animals, although a few are known from analysis of common poisoning incidents where the exposure of humans who died from exposure can be back-calculated. LD_{50} can vary by several orders of magnitude between species.
- **Gender.** The details of metabolism can differ between males and females, and one gender may be much more sensitive to a particular chemical.

- **Age.** Children are typically more sensitive to chemical exposures, because of their small size, differing (usually faster) metabolism, and the fact that a childhood exposure can act over a very long time.
- **Ethnicity.** Just as certain ethnic groups are known to be particularly vulnerable or resistant to particular diseases and disorders, toxins may affect members of different ethnicities differently. Such effects have received very little study.
- **Route of exposure.** Because, for example, the skin retards absorption of many chemicals, skin-absorption LD_{50}s are often much larger than ingestion LD_{50} values.
- **Presence of other chemicals.** The toxic dose of a compound can be affected by other compounds; chemicals dissolved in dimethylsulfoxide (DMSO), a common chemistry lab solvent, are carried easily through the skin—and thus toxic dermal absorption doses may be much lower than usual. Another example is certain medications (e.g., bupropion, an antidepressant also used for smoking cessation) that should not be taken with particular foods such as grapefruit or pomegranate due to the synergistic toxic effects observed with the foods' chemical constituents.

7.4.4.3 Exposure limits for toxic chemicals

Even though lethality and toxic-effect data are not useful to the researcher in his or her everyday work, one needs guidance regarding "how much is safe?" **Exposure limits** for chemicals provide that answer. Generally, exposure limits specify amounts of chemical that should not be exceeded in a given exposure scenario. For occupational exposures, this is usually chronic exposure over an 8 hour day for 40 years. Transient or short-term exposures above these levels may or may not be harmful—occupational exposure levels are not usually specified for acute exposure.

Exposure limits may be specified by governmental agencies, in which case they are generally binding on employers (including laboratories), or they may be recommendations made by professional bodies such as the American Council of Governmental Industrial Hygienists (ACGIH). These limits are computed based on debate among groups of toxicologists considering a broad variety of toxicity data, and the rigor of the proceedings may vary.

In the United States, the Occupational Health and Safety Administration of the US Department of Labor, with the assistance of the National Institute of Occupational Safety and Health (NIOSH), sets **Permissible Exposure Limits (PELs)** for a variety of chemicals; PELs are legally-binding limits on the amount of chemical to which a worker may be exposed over a certain time period. PELs are usually expressed as **time-weighted averages**, meaning, an average concentration or dose received over an 8-hour work shift (or sometimes a 15-minute time period—a **Short Term Exposure Limit, or STEL**); occasionally **Ceiling (PEL-C)** values, which are never to be exceeded are given. (Ceiling PELs are often specified for sensitizers such as TDI; see page 124 above.)

The regulatory process for issuing PELs is cumbersome and slow; most PELs have not been updated for decades. ACGIH also issues similar exposure limits called **Threshold Limit Values (TLVs)** for many more chemicals, based on a frequently-updated review process. TLVs are generally equal to or lower than the corresponding PEL, if a PEL exists, mostly because they are based on a broader variety of more recent data. TWA, STEL, and Ceiling TLVs are provided as appropriate to the chemical.

Similar processes exist in most countries. Smaller countries often adopt the exposure limits set by the US, Russia, or European Union, while amalgamations of countries (e.g, those of the British Commonwealth, MERCOSUR, or the EU) will share resources to develop common exposure limits.

The Vapor Hazard Ratio

The most common exposures to chemicals in laboratories are by inhalation or by skin absorption. The actual inhalation hazard offered by a chemical depends not only on the toxicity of the chemical but on the volatility: given the same toxicity, a volatile chemical is usually more easily transported into the lungs in higher doses and so offers a higher practical risk. The **Vapor Hazard Ratio (the VHR)** is a figure-of-merit that can be used to estimate the inhalation hazard presented by a particular chemical. The VHR is defined as the equilibrium vapor pressure at 20°C (68°F) divided by the Threshold Limit Value [72]. VHR calculations can be very useful in comparing inhalation hazards, for example, when choosing solvents, it is usually more prudent from a safety standpoint to choose the lower VHR solvent. Note that a higher-VHR solvent may be much more appropriate for the application, but comparing the VHRs allows one to judge whether the additional risk is justified. Various schemes exist [73] for determining the need for a chemical fume hood based on the VHR.

7.4.4.4 Work practices for highly-hazardous chemicals

The use of a chemical fume hood and appropriate protective gloves, lab coat, and chemical splash goggles usually suffice for safe handling of most types of toxic materials. When handling highly toxic materials, such as carcinogens, reproductive hazards, and other extremely-hazardous materials, additional measures are necessary. The Safety Data Sheet (see section 7.5.4 below) is the best source for required chemical-specific handling practices, but the list below gives some general guidelines.

Minimize the amount of chemical to be used. This means use the smallest suitable quantity for the experiment, but it also means **order the smallest size container available**. Removing 1 mL of chemical from a 2 L bottle is not only difficult, but it exposes the researcher to the risk of a much larger spill.

Double-gloving is one of the most common precautions taken when working with highly-hazardous chemicals. In its simplest form, double-gloving consists of wearing one pair of disposable gloves over another; the outer pair should be removed immediately upon any hand exposure to the chemical (e.g., dripped material), hands washed, the inner pair removed, hands washed, and both pair of gloves replaced. It should be noted that if the chemical penetrates the glove quickly enough, this form of double-gloving is insufficient to protect the researcher—it would not have helped Karen Wetterhahn, for example. (See section 1.1.2 above.)

A more protective form of double-gloving is used for extremely-hazardous compounds or those for which the penetration time is not known and cannot be estimated. In this style of double-gloving, an outer pair of gloves believed likely to be resistant to the chemical in use is worn **over top** of an inner glove or glove liner. Often the inner glove is a SilverShield® (North) or similar multilayer polymer sandwich. (See page 121.) Note that this practice is only viable using thicker reusable gloves, and so the concomitant loss in dexterity should be included in work planning.

Spills should be cleaned up immediately. While this is a good practice for use of any chemicals, it is essential in working with highly-toxic materials, as it prevents spread of contamination throughout the laboratory. Disposable absorbent pads may be used (with impermeable side **down**) underneath work areas. Use two pads, because in the event of a small spill, the upper pad can be disposed and work can continue quickly.

Check the work area for spills immediately after finishing work, so that the spill does not have time to dry. This makes the spilled material easier to find and prevents it from becoming airborne. The author uses a small ultraviolet LED flashlight to search for spills in laboratories using fluorescent materials such as ethidium bromide.

In the case of extremely dangerous materials, particularly if they are used in relatively large quantity, a **glove box** should be used, kept under negative pressure relative to the lab. The glove box may require special filtration or treatment of the exhaust, depending on local regulations and the toxicity properties of the material. As with regular gloved operations, appropriate materials for the gloves must be chosen. In extreme cases, special-purpose facilities providing **remote operation** may be necessary—this is sometimes done with extremely radioactive samples, for example. (See Figure 9.1 below.)

7.4.4.5 Storage of highly-toxic materials

Highly-toxic materials should be stored in a locked cabinet in a locked lab because of the risks associated with unauthorized personnel gaining access. For pharmaceutical materials, some of which can be highly toxic (e.g., cancer drugs), this is done as a matter of course, required by various drug-control regulations. Actual storage methods depend on the physical state of the chemical.
- **Solid** toxics may be stored in a cabinet. If they are water-reactive, a small desiccator may be placed in the cabinet.
- **Liquid** toxics should be stored with secondary containment. Note that secondary containment should always have at least the volume of the bottles to be contained to prevent overflow in the event of a leak. Secondary containment could

be a plastic tray or simply a large plastic beaker; do not use glass for secondary containment except for highly-oxidizing materials that would react with plastics.
- **Gaseous** toxics can be stored in specialized ventilated **gas cabinets**, often available from the gas supplier. For highly-toxic gases such as arsine (AsH_3), fire codes often require gas cabinets.

7.4.5 Physical hazards from chemicals

Despite the focus on chemical hazards, chemicals also present physical hazards; these are often underestimated.
- **Chemical explosions.** Uncontrolled reactions frequently result in detonation or deflagration of a bulk material. The force of these explosions, particularly if the material detonates, can be extremely destructive.
- **Overpressure.** Increase in number (a reaction that generates more molecules than it started with) or temperature in a closed vessel often produces overpressure events. If relief capacity is insufficient, a pressure explosion results as the vessel ruptures.
- **Detempering or apparatus flaws.** Excessive temperature may weaken a reactor or container's materials of construction—metals, for example, may lose their "temper" and become drastically weakened. Metal structural elements of buildings are often insulated with fire-resistant materials in order to prevent fires from detempering the steel and causing the building to collapse. Similarly, many materials lose strength or become vulnerable to impact at low temperatures—consider the brittleness of common steels held below −28°C (−20°F) or that of polymers used below their glass transition temperatures.

7.4.6 Reactive chemicals

"Reactive chemicals" is an overarching term for any chemical or chemical combination that has dangerous properties not clearly identifiable as fire-related, corrosive, or toxic. Materials may be reactive by themselves or may be strongly reactive with each other—particularly with water.

7.4.6.1 Inadvertent reactions
Inadvertent chemical reactions are the most common type of "reactive chemical" incident. They can present a hazard in several ways:
- **During experiments.** Energetic side reactions occasionally occur during chemical experimentation, causing unexpected exotherms (heat excursions) or an unexpected increase in the number of molecules present (which in gas-phase

reactions causes a large pressure rise in the container). Unexpected reactions may produce "energetic materials," a euphemism for explosives. Researchers conducting so-called "one-pot" or "green chemistry" experiments may be particularly vulnerable to this hazard since multiple reactions are performed in series without intermediate purification.
- **In waste disposal.** Chemicals are as reactive in the waste bottle as in the reagent bottle. "Waste" is not a magical safe material. Chemical reactions in waste have caused many accidents, oftentimes causing the container to explode due to overpressure. An example occurred recently at a US Department of Energy radioactive waste disposal facility, where a scientist inadvertently used "organic" (made from wheat rather than clay) cat litter in a drum of waste to absorb spills. The drum mainly contained waste nitrate salts, which are moderately strong oxidizers; mixing them intimately with organic materials (fuel) created the conditions for an explosion. (See Figure 7.5.)
- **In storage.** Aging, light or heat exposure, or exposure to atmosphere can all produce reactions or reactive by-products. For example, some organic materials form sensitive explosive peroxides upon exposure to air or oxygen.
- **Self-reactivity.** Some compounds, particularly those that contain oxidizing or high-energy groups (nitrates, azides) can undergo intramolecular reactions that release energy. If this property is not anticipated, an unanticipated reaction may occur.

Figure 7.5: Aftermath of explosion in nuclear Waste Isolation Pilot Plant, Carlsbad, NM (US). [74] Photo courtesy U.S. Department of Energy.

7.4.6.2 Intentional reactions

Reactive chemical hazards may also exist in **intentional chemical reactions,** particularly if the reactive hazards are not adequately controlled.
- **Exothermic reactions.** Reactions with significant energy release, particularly if the reaction rate is fast or uncontrolled, can overheat or explode. Polymerization reactions are particularly prone to this because of the Trommsdorff-Norrish effect [75].
- **Increase in number.** Some reactions produce more molecules than they consume—dissociation or depolymerization reactions being the classic example. Particularly in gas-phase reactions, this often leads to an increase in pressure and/or volume. If not adequately allowed for in the design, the pressure increase can rupture the reaction container quite violently.
- **Gas-generating reactions.** Some reactions generate gases as a by-product (or the intended product). Even if the number of molecules is the same on both sides of the chemical equation, the gas pressure exerted on the vessel can be quite high.
- Any of the above classes of chemical reaction conducted in a sealed vessel or other system **must have adequate pressure-relief capacity.** Pressure relief design for reacting systems is a specialist field using specialist methods [51]; consult a qualified professional for assistance if necessary. Pressure relief devices must discharge to a safe location.

7.4.6.3 Important classes of reactives

There are several common types of reactive chemical hazards in the laboratory. This list is not exhaustive.
- **Explosives, explosive mixtures, and other "energetic materials."** Explosives typically undergo intramolecular reactions between oxidizing and fuel groups; the classic example is trinitrotoluene (TNT). The incident involving Preston Brown (see section 1.1.5 above) involved nickel hydrazine perchlorate, a salt containing both fuel (hydrazine) and oxidizer (perchlorate) groups.
- **Monomers.** Many monomers or prepolymers are prone to rapid or explosive polymerization, particularly in the presence of contaminants such as oxygen, peroxides (including their own peroxidation products), and metals & other polymerization initiators.
- **Azides and other polymerization initiators.** Many of these compounds are designed to disintegrate easily and require refrigerated storage. Some require special transport and safeguarding to prevent explosions in transit.
- **Nitrates and other nitro compounds.** Nitrates tend to be oxidizers and can cause gas generation or explosive heat release, particularly when mixed intimately with a fuel. Ammonium nitrate has caused several large-scale disasters, including the near-complete destruction of the port of Texas City, TX (US) in 1947

[76], both by itself and in combination with organic materials. Ammonium nitrate mixed **in situ** with fuel oil (or diesel fuel)—often called "ANFO"—is a high-volume explosive in the mining industry.
- **Perchlorates.** Perchlorates are notoriously self-reactive, as well as highly oxidizing in many cases. Ammonium perchlorate is used as a constituent of solid rocket fuels; detonation of the ammonium perchlorate plant in Henderson, NV (US) on 4 May 1988 demonstrates the amount of energy that can be released by perchlorate materials [77]. Videos of this event may be easily found on the Internet.
- **Nital and other metallurgical etchants.** Many mixtures used in metallurgy, such as "nital", an ethanol-nitric acid mixture, can be highly hazardous if not properly made, used, stored, and disposed [78].

7.5 Communicating chemical hazards

Because the chemical hazards of a material can be very complex, several ways have been developed to communicate those hazards. These systems are usually customized toward particular applications, such as emergency response.

7.5.1 NFPA 704 "fire diamond"

One of the most well-known chemical hazards communication systems is designed for warning emergency responders of the gross hazards of a material. In emergency response situations, there are three major hazardous chemical properties of concern:
- **Does the material burn, and if so, how flammable is it?** Fire is of immediate interest to firefighters called to the scene of a chemical incident.
- **Is the material toxic, particularly in an acute manner?** Beyond the obvious concern for the responders, an incident commander must determine quickly whether any evacuation of the area is necessary or if there is danger to responders from contact with the material.
- **Is the material chemically-reactive?** Reactive chemicals such as ammonium nitrate and calcium carbide have been responsible for large-scale incidents, such as the ammonium nitrate fertilizer explosion referenced above.

A simple 0–4 rating of these factors is the basis of the **NFPA 704 system** (NFPA: National Fire Protection Association) [79]. (See Figure 7.6.) The first number, with a blue background, rates the health hazards of the material, from 0 for almost harmless to 4 for potentially lethal. (There are technical toxicological standards for each category.) The second number, with a red background, communicates the fire and combustion explosion hazards: again 0 for no significant hazard and 4 for extremely flammable. The third number relates the material's reactivity, ranging from 0 for nonreactive to 4 for "unstable." A fourth space is provided for special information, such as "use no water" for water-reactive compounds like sodium metal.

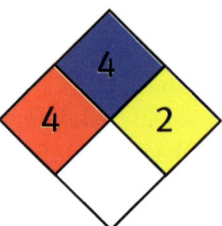

Figure 7.6: Typical NFPA 704 label for phosgene.

"NFPA 704 Diamonds" are often seen on storage tanks and other fixed facilities in the US to which emergency responders might be called. Other countries use similar systems.

The Hazardous Materials Information System (HMIS), originally a coatings-industry standard, is very similar to NFPA 704, with additional advice on personal protection and an improved focus on chemical use as opposed to emergency response. While it can be useful for understanding laboratory risks from a chemical, it has largely been superseded by the Globally Harmonized System (GHS; see section 7.5.3 below).

7.5.2 Transportation labeling

While NFPA 704 is usually used for fixed facilities emergency response, chemicals in transport use a different system: colored and numbered placards affixed to the transport vehicle and package. Transport labels are meant to indicate the most important information about a chemical and to allow its identification from a distance in the event of a serious transport mishap, such as a tanker truck crash that results in a leak of the truck's contents.

Each label (there are about 25 in use) contains a number announcing the **principal hazard class** such as "flammable liquid" or "corrosive," as well as a pictogram further identifying the type of hazard. Most hazard classes have two or more divisions characterized by the degree of hazard. One or more labels may be used for a given chemical. Figure 7.7 shows the labels typically used for fluorine in the US. There can be some differences in labeling among countries, (e.g., fluorine uses Poison Gas 2.3, Oxidizer 5.1, and Corrosive 8 labels in Europe). An accompanying chemical shipping identification number (standardized by the United Nations) is placed on the package; emergency responders typically use this number as an index into a book of standard protocols.

Figure 7.7: US Department of Transportation hazardous materials labels for fluorine.

7.5.3 The Globally Harmonized System

More useful in the laboratory is the **Globally Harmonized System** of chemical hazard communications [80]. GHS, as it is universally called, is a complex international standard method of labeling and communicating the hazards of a material **as they apply in the workplace.** Whereas NFPA 704 and the transportation labels attempt to relate only the principal hazards of the material, GHS aims to be as comprehensive as possible.

GHS consists of several main parts:

- A set of pictograms indicating the **types** of hazard present. Examples would be "Chronic health hazard," or "Oxidizer." Acetylene receives "Flame" and "Compressed gas" pictograms.
- An alphanumeric category giving more detail on the hazard types. **Category 1 flammable gases** are flammable in air at atmospheric temperature and pressure either a) in mixtures of less than 13% by volume with air or b) have a flammable range of at least 12% between upper and lower explosive limit. Acetylene is a Category 1 flammable gas. Figure 7.8 shows some of the pictograms used.
- A **signal word** (either "Danger" or "Warning") indicating the severity of the hazard. Category 1 flammable gases such as acetylene receive a "Danger."
- One or more **hazard statements** that give a short summary of the hazard. In the case of acetylene, the hazard statements would be "**H220: Extremely flammable gas,**" and "**H280: Contains gas under pressure; may explode if heated.**" The hazard statements are numbered to facilitate understanding in different languages.
- One or more **precautionary statements** indicating to the user what practices should be followed to minimize risk from the chemical. One of the hazard statements for acetylene reads, "**P210: Keep away from heat/sparks/open flames/hot surfaces—No smoking.**" The precautionary statements are numbered for the same internationalization reasons as hazard statements.

The United Nation's GHS specification comes with a companion "model regulation" to encourage countries to adopt the specification without significant modification.

Generally, the GHS pictogram, category, and signal word may be found on any properly-labeled chemical bottle produced in the last several years; depending on the container size, the hazard and precautionary statements may or may not be included. These statements, as well as a plethora of other safety-related information, may be found on the chemical's Safety Data Sheet, which always accompanies chemicals in transit.

Figure 7.8: Example Globally Harmonized System pictograms.

7.5.4 The Safety Data Sheet

The **Safety Data Sheet (SDS)** is a critical tool for laboratory chemical users, and it is an integral part of the GHS system. The SDS in theory provides "comprehensive information about a substance or mixture for use in workplace chemical control regulatory frameworks. Both employers and workers use it as a source of information about hazards, including environmental hazards, and to obtain advice on safety precautions" [80]. In the United States and in certain other countries, the SDS was formerly known as the "Material Safety Data Sheet (MSDS)."

SDSs have a standardized format, beginning with the identification of the material and its hazards and progressing to further and further detail about those hazards and appropriate handling information. There are sixteen parts [80]:

1. Identification;
2. Hazard(s) identification;
3. Composition information on ingredients;
4. First-aid measures;
5. Fire-fighting measures;
6. Accidental release measures;
7. Handling and storage;
8. Exposure controls/personal protection;
9. Physical and chemical properties;
10. Stability and reactivity;
11. Toxicological information;
12. Ecological information;
13. Disposal considerations;
14. Transport information;
15. Regulatory information;
16. Other information.

Note the use of the phrase "in theory" in the second sentence of this section. The Safety Data Sheet is used for a number of purposes, including as a transport hazards document and a workplace hazards document for a variety of types of workplaces. The same SDS is used for responding to a spill of a truckload of chlorine as for the use of a large (1-ton) cylinder of chlorine in a water treatment plant; it will also be employed for laboratory use of a single small cylinder.

This "all things to all people" approach can make it difficult for laboratory users to find the information most applicable to the lab. Spill information may be written for large-scale transport spills; the author has seen an SDS for table sugar directing that cleanup of a spill should only be done in a sealed chemical suit with water mist to suppress dust. While sensible for a railcar-sized spill, where the dust can be explosive, such directions are ridiculous in the lab.

SDSs are also heavily influenced by legal considerations and tend to be excessively conservative. Oftentimes, directions for emergency response and even general use of a material may drastically overstate sensible precautions. Another SDS in the author's possession instructs persons exposed to water to rinse the affected area for 15 minutes—with water.

Quality problems notwithstanding, laboratory users should **read and understand** the SDS **before beginning work with the chemical.** SDSs contain information that may be critical to the safety of the experiments undertaken; not noting relevant information can be the root cause of an incident. In one incident, a researcher neglected to read the SDS provided with a catalyst purchased from a major supplier, which plainly stated that the material was capable of igniting flammable solvents if allowed to partially dry. During his experiment, the researcher abandoned the catalyst after filtering it, where it partially dried and ignited the methanol solvent used; the resulting fire did $50,000 damage to the building.

7.6 Case studies

7.6.1 Chemistry experiment

In a chemistry laboratory, researchers synthesize and study **oligosilanes**, that is, short-chain analogues to typical organic chemicals in which the carbon chain has been replaced with silicon. The semiconducting properties of bulk silicon translate into unusual electronic properties, as certain electronic orbitals (σ molecular orbitals) delocalize over the chain. This delocalization provides an easy path for electron transport across the length of the molecule, and it makes these low-toxicity compounds ideal components for nonlinear optoelectronic devices.

One particular study in the subject laboratory involved combining standard oligosilanes with various organic groups that exhibit a different type of molecular orbital delocalization (π-conjugation) to study whether such hybrid functionalized materials might provide more dramatic properties than the standard materials. In fact, they do. In [81], investigators determined that oligosilanes with cyanovinyl end groups, in particular, exhibited high charge-carrier mobility (a measure of the ability of electrons or missing-electron 'holes' to move through the material) and other electronic properties that analogous molecules with a carbon backbone in place of the silicon did not. Particularly interesting is the fact that the charge carrier mobility in the new materials was 10–100 times that of previously-synthesized oligosilanes.

The Supplementary Information to [81] provides complete details of the synthesis of the oligosilanes and organic analogues studied in the paper. Like the syntheses of most organic compounds, chemical hazards abound in the synthetic procedures. Two types of chemical hazard present themselves obviously to the reader: the use of toxic & flammable solvents, and the use of active metal compounds.

Procedures for making "Si_6C_6" (see Figure 7.9 for structures and synthetic scheme) and intermediate compounds in the synthesis involve use of benzene, ethyl acetate, hexanes, N,N-dimethylformamide (DMF), diethyl ether, and piperidine (hexahydropyridine). Ethyl acetate is likely the least hazardous of the above, but it is highly flammable (as are all of the mentioned compounds). Benzene, besides being flammable (it is a component of gasoline) is a known human carcinogen, causing blood cancers after long exposure at relatively low levels. Dimethylformamide and piperidine are moderately toxic, and hexanes are neurotoxic.

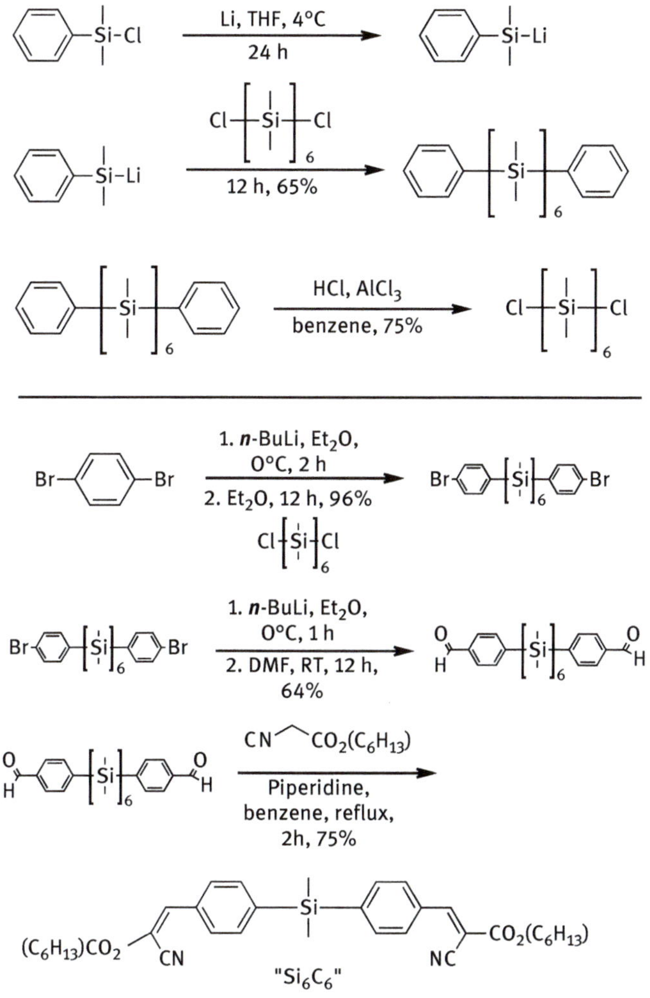

Figure 7.9: Structures and synthetic scheme from [81].

DMF carries the extra complication of being a reproductive toxin, an item of particular concern in a laboratory whose researchers are of childbearing age. (Recall also that reproductive toxins are not limited to effects upon females; some compounds cause damage to male germ cells.) Fortunately, DMF's ability to cause chromosomal abnormalities in humans is limited, and in animals, reproductive toxicity has been observed only at exposure levels much greater than those which cause liver toxicity [82].

Several of the steps in the manufacture of "Si_6C_6" also involve the use of active metal compounds. While some of the other syntheses in the paper employ the raw metals themselves, particularly lithium, the creation of "Si_6C_6" uses **n**-butyllithium (**n**BuLi; $LiCH_2CH_2CH_2CH_3$), a "superbase" capable of removing protons from locations no other base can access, and (most useful in this case) a very strong nucleophile (a chemical capable of displacing other groups such as halogens from organic compounds). **n**BuLi, like the **t**BuLi involved in the death of Sheri Sangi (see section 1.1.6 above), is extremely flammable and pyrophoric. In this case, the nBuLi assists in substituting a cyanovinyl group for a bromine atom at a particular position in the molecule.

The lab uses a variety of strategies to control risk from the plethora of chemical hazards listed above. Aside from the use of standard laboratory personal protective equipment (including use of flame-retardant lab coats when handling pyrophoric materials or active metals), two techniques are of particular use: work at small scale and work in sealed systems.

Since many of the compounds used and created in the laboratory are air-sensitive or moisture-sensitive, it is standard procedure to handle them under nitrogen or argon atmosphere. A device called a **Schlenk manifold** (or "Schlenk line") facilitates this process by providing a ready source of inert gas and a means to interconnect containers without exposing the contents to air. In Figure 7.1, the Schlenk manifold is readily visible mounted at the back of the hood. Materials are transferred in and out of the sealed inert-atmosphere system with syringes inserted through rubber septa. In addition to providing "product protection," this system provides a significant degree of personnel protection by preventing exposure of pyrophoric materials to air and release of toxic or flammable vapors into the laboratory atmosphere.

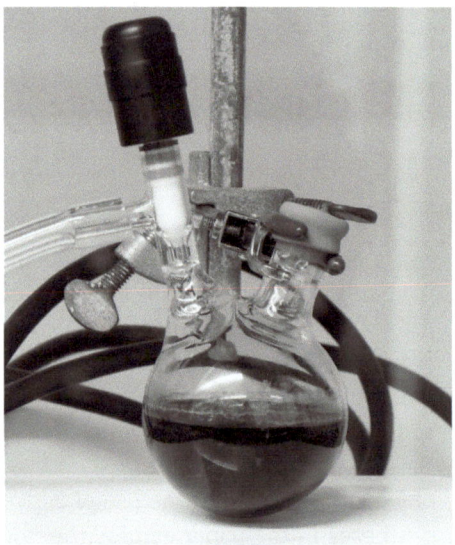

Figure 7.10: Gram-scale reaction involving elemental sodium.

The other essential practice used in this laboratory is to work at the smallest scale consistent with obtaining the necessary quantity of product. While many chemistry laboratories employ scales of tens of grams, this laboratory typically works on the scale of 1 gram or so. The final step in synthesis of "Si_6C_6", for example, involves use of 1.5 mL of **n**BuLi solution—recall that Sheri Sangji died while attempting to withdraw over 50 mL of **t**BuLi solution (and that was but a third of her eventual planned use). In some cases, syntheses can be performed at the milligram scale, but as this laboratory requires macroscopic quantities of product to make electronic devices for testing, this is impractical. Figure 7.10 illustrates a typical metallation reaction being run on the scale of several grams, in this case using elemental sodium as the active metal.

7.6.2 Biology experiment

A biology laboratory studies clathrin-mediated endocytosis—the importation of target molecules into a cell— in budding yeast (**S. cerevisae**) using various molecular biology techniques such as protein labeling and gene knockouts [83]. The process of endocytosis is thought by some to begin when an "adaptor protein" such as Syp1 is recruited to the cell membrane surface as the concentration of a cargo molecule reaches a critical value. Adaptor proteins respond to various cellular sorting signals such as ubiquitin or other peptide chains on the surface of the cargo to be imported.

The adapter protein recruits **clathrins**, 6-chain proteins that self-assemble to form various types of curved latticeworks [84]. These latticeworks help create a **vesicle**, a pocket into which the molecules to be imported are drawn. Enzymes (**actin polymerases**) then grow **actin** fibrils, long-chain structural proteins, which pull the pocket away from the surface of the cell; the clathrin-lined pocket is then cleaved from the cell membrane surface, forming an intracellular vesicle in a process called **scission**. As the actin fibrils extend, the now-isolated vesicle is pushed into the cell and may then be delivered to intracellular compartments called **endosomes**.

In a typical experiment, the researcher aims to study the motion of an adaptor protein on the surface of the cell membrane. One way to accomplish this study is to "tag" the adaptor protein with **green fluorescent protein** (GFP), a short (238-amino acid) protein obtained originally from the jellyfish **Aequorea victoria** [85]. This protein fluoresces bright green when illuminated with blue or ultraviolet light, enabling visualization of its position.

If the gene for the adaptor protein can be isolated using standard molecular biology techniques, it can be modified using recombinant DNA technology, linking it to a GFP "tag." Once the gene is isolated and its DNA sequence determined, the researcher synthesizes a **plasmid** (circular DNA fragment) containing the gene of interest, including the coding sequence for green fluorescent protein. The plasmid is then transferred (the technical term is **transformed**) into an intermediate host organism such as **E. coli** for amplification (production in quantity), isolated, and transformed into the yeast cell for study.

The end product is a plasmid instructing the yeast to synthesize a version of the adaptor protein with GFP attached to one end. The amount of the modified protein expressed can be controlled by various factors such as the number of plasmids per yeast cell. Once the labeled protein is present in an adequate concentration, the researcher can follow its position at the cell membrane using fluorescence microscopy and use that data to study the process of endocytosis.

As a confirmatory quality-control step after creating the plasmid and before amplification, it is prudent to verify that the plasmid made actually contains the genes of interest. This is accomplished by **"DNA fingerprinting"** (gel electrophoresis) technology, in which a sample of plasmid is cut into pieces by a **restriction enzyme** and the size of the pieces measured by electrophoresis. In electrophoresis, the restricted sample is injected into a gel, typically of agar-agar or acrylamide; acrylamide, a toxin considered to be a potential human carcinogen, is more commonly used for separation of proteins than DNA.

Upon application of a strong electric field (several V/cm), the cut DNA fragments, which carry a pH-dependent charge from their sugar backbone, move in the direction of the field at a speed inversely proportional to their mass. The DNA is then made visible by staining—and this step potentially involves a serious chemical hazard. The most common DNA stain used is **ethidium bromide** (EtBr), a fluorescent molecule which intercalates itself between the DNA strands. The gel is treated with a solution of EtBr and illuminated, usually from beneath, with a strong ultraviolet light, emitting a characteristic orange fluorescence which can be photographed or scanned as a permanent record.

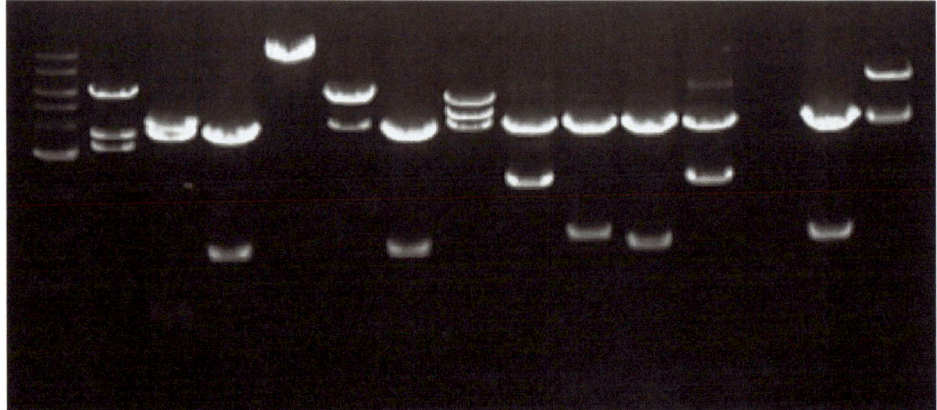

Figure 7.11: Gel electrophoresis scan with Midori Green. Photo courtesy Kyle Hoban.

Unfortunately, EtBr is a mutagen, causing DNA changes in the standard Ames test; interestingly, EtBr itself is not thought to be mutagenic, but its metabolites (liver breakdown products) are [86]. More detailed investigative work has shown that EtBr is not a strong mutagen in humans but can be toxic at high concentrations, working in a manner similar to several human anticancer (chemotherapeutic) drugs. Disposal of EtBr should be done in a manner similar to that of other chemotherapeutic drugs. At the author's institution, ethidium bromide solutions are disposed by filtration through activated charcoal, which absorbs the EtBr; the activated charcoal is then itself disposed as hazardous waste or by incineration as chemotherapeutic waste.

Due to the potential toxicity and disposal headaches brought on by the use of ethidium bromide (as well as the non-ionizing radiation hazard of the UV transilluminators used to visualize EtBr-stained gels; see section 9.2.1 below), this laboratory has changed visualization agents from ethidium bromide to Midori Green (MG). MG is known not to be carcinogenic, and standard glove types used in molecular biology (such as disposable nitrile gloves) are resistant to penetration by the agent. It also fluoresces under illumination with blue/green light, allowing use of a LED transilluminator instead of the more-hazardous UV units. (See Figure 7.11.) It breaks down quickly in the environment, simplifying disposal considerably. It is considerably less toxic than EtBr (oral LD_{50} for Midori Green is greater than 10 g/kg, [87], while the oral LD_{50} for ethidium bromide is approximately 1 g/kg and the LC_{50} inhalation toxicity is on the order of tens of parts-per-billion [88]). (See Figure 7.12.)

Figure 7.12: Electrophoresis apparatus with label.

The **substitution** of Midori Green for ethidium bromide is a classic example of the application of Inherently Safer Design (see section 15.3.2 below) to create an experiment that completely avoids a hazard and thus avoids the need to undertake special measures to control it.

7.7 Exercises: Chemical hazards

1. Answer the following questions:
   A. True or false? A chemical fume hood should be used for handling any volatile chemical.
   B. True or false? A fire due to flammable materials may be avoided by excluding nitrogen from the area.
   C. The ratio of vapor pressure to a substance's threshold limit value is called what?
   D. What type of eye protection should be used for most chemicals?
   E. True or False? Once used, all waste and excess chemicals from an experiment should be placed in a single glass bottle for disposal.
2. Using any source of Safety Data Sheets available to you, answer the following questions. Cite your source.
   A. What is the vapor pressure of hexane at 25°C (298 K, 77°F)?
   B. Is acetone toxic?
   C. What are the principal hazards of sulfuric acid (fuming)?
   D. What is the appropriate first aid for exposure to hydrofluoric acid (50%)?
3. Examine the following brief laboratory scenarios, identify chemicals present and their related hazards, and give at least one engineering or administrative control measure suitable to reduce risk from those hazards.
   A. Zinc metal is dissolved in hydrochloric acid to make zinc dichloride.
   B. A silicon wafer is etched by immersion in a bath of hot hydrofluoric acid (50%), followed by rinsing and immersion in a bath of hot Piranha solution (concentrated sulfuric acid and concentrated hydrogen peroxide) to remove a photoresist applied to the surface to control the etching.
   C. As a step in the synthesis of an organic compound, an intermediate product is reacted with **tert**-butyllithium in pentane, withdrawn by syringe from a stock bottle. The synthesis continues by addition of methyl iodide.
   D. A cleaning solution is made from 50% sulfuric acid, 1% ammonium bifluoride, and balance water.

8 Biological hazards

Biological hazards (biohazards, for short), generally refer to hazards posed by living organisms, their constituents, or their products. **E. coli**, the common intestinal bacterium, deoxyribonucleic acid (particularly if processed via recombinant DNA processes), and biologically-produced toxins such as aflatoxin (a strong carcinogen produced by certain types of mold that grow on improperly-stored peanuts) all qualify as biohazards.

In this chapter, we will briefly address the nature of biohazards and some simple laboratory practices used to maintain a safe work environment. The broad range and complexity of living organisms means that biological safety is both simpler and more complex than, for example, chemical safety. A researcher can work with most classes of hazardous chemicals without professional assistance (in at least simple use scenarios), but the author does not recommend working with biological materials without the advice of and appropriate training from a biosafety professional.

8.1 Lab-acquired infections

The principal concern with laboratory work involving living organisms such as bacteria (or "semi-living" organisms such as viruses) is **laboratory-acquired infection (LAI)**. Over the past few decades, the incidence of **reported** LAIs has decreased significantly in the United States, despite the explosion in biological research. This reduction may be real, but it is thought by some to be due to concealment of the infection, either by researchers who do not wish to be censured for improper handling of biological materials or by employers who desire to avoid the reputational damage (and possible lawsuits) resulting from a biological incident. Incidents which are more difficult to conceal (e.g., deadly LAIs or mishandling of biological samples that leave the lab) and garner more press coverage.

An example of this would be the inadvertent shipping by the US Centers for Disease Control and Prevention (CDC) of a highly-infectious avian flu virus transmissible to humans—this incident drew attention from the national press and prompted a formal CDC investigation [89] During the resulting CDC-wide "Safety Stand-Down" to search for unauthorized or unidentified biological agents, approximately 31 samples of highly biohazardous "Select Agents" were found in unapproved CDC labs.

The range of laboratory organisms that have led to laboratory-acquired infections varies widely. It includes such organisms as **Mycobacterium tuberculosis** (the bacterium which causes tuberculosis), **Coccidioides** species (fungi which are commonly associated with archaeological digs, particularly in the American Southwest), **Plasmodium** species (which can cause malaria among researchers working with it), or **Herpesvirus simiae** (Monkey "B virus," a deadly infection carried by macaque monkeys—which are commonly used in certain research fields). Many other microorganisms can cause LAIs, particularly if the victim is immunocompromised or exposed via direct blood exposure (such as a needlestick with contaminated body fluids). A more comprehensive listing of laboratory pathogens may be found in [90].

8.2 Assessment of biological infection risk

Before working with any biological material, it is essential to assess the risk of exposure and consequent infection. Biosafety professionals consider a variety of factors in determining the hazard level of a biological experiment. These factors divide into two basic categories: inherent hazards of the biological agent (**agent hazards**) and hazard scenarios introduced by the procedures to be used (**laboratory procedure hazards**) [90].

8.2.1 Agent hazards

First and foremost, the most important factor in determining the biological risk of an organism is the **pathogenicity** of the agent, that is, its capacity to cause disease or other undesirable biological harm. Note that many agents are essentially non-pathogenic to healthy humans but may be deadly to those with suppressed immune systems (e.g., those taking immunosuppressive drugs after a transplant operation or for conditions such as lupus, or persons with active human immunodeficiency virus—HIV—infection). The easier the infection and/or the more serious the resulting disease, the more probable and/or severe is the risk.

Many agents may be pathogenic to nonhuman hosts, such as hoof-and-mouth disease in cattle, or may be **zoonoses.** Zoonotic organisms are primarily hosted in one species, sometimes with little ill effect, but can be transmitted to other species—the monkey B virus is often found in macaque monkeys, who usually show few or no symptoms [91], but the virus can be easily transmitted to humans through a monkey bite, producing severe brain damage and/or death if not immediately treated.

Hand-in-hand with the pathogenicity of the organism is the **availability of vulnerable hosts**. Even a deadly organism may be used in a relatively low-risk environment if its host is not available and it is not easily carried from place to place. Studying a bacterium that causes a deadly disease in snow leopards (found only in the Himalayas or in zoos) **in vitro** in a laboratory in Mexico, while posing a severe hazard, is highly unlikely to be transmitted and thus is a relatively low-risk biological experiment. Handling hoof-and-mouth disease in a cattle-raising region is high risk and requires special risk-reduction measures, including at times handling at a special high-containment agricultural bio-lab. (See section 8.3.4 below.)

The organism's **transmissibility and mode of transmission** (or route of infection) dramatically affects the probability of a laboratory-acquired infection. Viruses which require direct contact with body fluids, such as HIV, are often handled under BSL–2 procedures in BSL–2 laboratories (BSL=Biosafety Level; see section 8.3.2 below), at least for smaller quantities handled in a manner that minimizes aerosolization. **Mycobacterium tuberculosis**, on the other hand, is readily spread by inhalation, and is usually handled at BSL–3 (see section 8.3.3 below).

Another factor in the inherent risk of an organism is the **detectability and treatability of the resulting disease**. Obviously, an organism such as the mite **Sarcoptes scabiei**, which causes the itchy and visually obvious condition called scabies (a human form of mange), and which is easily treated with the acaricide (mite-killer) permethrin, is a much lower risk than **Ebolavirus sp.**, the causative agent for Ebola hemorrhagic fever, which is currently untreatable.

Table 8.1: WHO Risk Groups for infective microorganisms

Risk Group	Risk to individual	Risk to community	Description
1	No/low	No/low	A microorganism that is unlikely to cause animal or human disease.
2	Moderate	Low	A pathogen that can cause human or animal disease but is unlikely to be a serious hazard to laboratory workers, the community, livestock, or the environment. Laboratory exposures may cause serious infection, but effective treatment and preventive measures are available and the risk of spread of infection is limited.
3	High	Low	A pathogen that usually causes serious human or animal disease but does not ordinarily spread from one infected individual to another. Effective treatment and preventive measures are available.
4	High	High	A pathogen that usually causes serious harm or animal disease and that can be readily transmitted from one individual to another, directly or indirectly. Effective treatment and preventive measures are not usually available.

Source: Table 1 of [92].

The World Health Organization (WHO) classifies pathogens into risk groups using (mostly) the factors listed above. Table 8.1 shows the WHO definitions of biological risk groups. Note that risk group does not necessarily equate to the Biosafety Level appropriate for a given experiment. Biosafety professionals take note of the WHO Risk Group of an organism when determining precautions to be taken with that organism.

Ease of sterilization is relevant when determining biological risk as well. An organism that would otherwise be studied under BSL–1 (see section 8.3.1 below) or BSL–2 conditions might require laboratory practices more appropriate to BSL–2 or

BSL—3 if the organism is difficult to kill. Agent **stability in the environment** is a related factor.

Lastly, organisms containing **recombinant DNA** are presumed more risky than the natural species, because of the possibility of inadvertent increase in disease severity or transmissibility. Recently, organisms containing **synthetic DNA** (from small amounts of added synthetic material to wholly synthesized viruses) have been given the same consideration.

8.2.2 Laboratory procedure hazards

Laboratory procedure hazards depend more on the way in which the organism is used or studied, as opposed to inherent properties of the agent. Probably the most important laboratory procedure hazard is the **skill level and care used by the investigators**.

Biological research depends upon constant attention to proper procedures and practices in the laboratory. Not only does careless work contribute strongly to the risk of laboratory-acquired infection, but it also threatens the integrity of the research results more seriously than in many other fields. In a chemical synthesis, cross-contamination of one batch of chemicals with another is tolerable, either because the contamination level is low enough to not disrupt the overall reaction or because the contamination will be removed in a later purification step. In biological work, a single cell landing in the wrong culture flask can thoroughly confuse results or even kill the organism of interest, substituting itself for the desired culture.

As one author stated, "Training, experience, knowledge of the agent and procedure hazards, good habits, caution, attentiveness, and concern for the health of coworkers are prerequisites for a laboratory staff in order to reduce the inherent risks that attend work with hazardous agents. Not all workers who join a laboratory staff will have these prerequisite traits even though they may possess excellent scientific credentials" [93].

Note that skill level and the quality and consistency of procedures do not necessarily go hand-in-hand: an expert investigator may have such sloppy work habits that others are unwilling to collaborate with him or her—or even to allow him or her into the lab! Conversely, a newly-hired laboratory assistant, if properly trained, is often capable of carrying out established procedures with precision and accuracy—but if any unusual situations arise, the technician may lack the expertise to adapt to the situation in a manner that maintains safety and scientific integrity.

Another important procedural factor in establishing appropriate biosafety precautions is the **quantity of material** to be used—all other things being equal, more agent equates to more risk. Thus, handling of HIV, usually done at BSL—2, might be restricted to a BSL—3 laboratory if the purpose of the work is to create large batches of virus for distribution to other researchers.

Finally, the **use modality** (the way in which the experiment is to be conducted) influences biological risk strongly. Laboratory protocols with a high potential for producing inhalable biological materials (biological aerosols), such as centrifugation, sonication, or simple shaking of open containers naturally pose a higher risk than other experiments. The possibility of injection of samples by sharps (needles, scalpels) is also taken into consideration.

8.3 Biosafety levels

Because of the variation in hazard levels presented by different biological agents, the degree of containment necessary to protect laboratory personnel, the public, and the environment varies as well. In the interest of standardization (and because many agents require similar precautions and laboratory facilities), four containment levels, or **Biosafety Levels**, abbreviated BSL—1 through BSL—4, are defined. Note that BSL designations apply both to the laboratory itself (its construction and equipment) as well as the practices used therein. Special measures for animal laboratories (agricultural biosafety levels) exist, and facilities involving research on animals are rated with an ABSL as well as a BSL [92].

8.3.1 Biosafety level 1 (BSL—1)

Biosafety Level 1 is the lowest level of containment, used typically for organisms deemed unlikely to cause disease or spread from the laboratory. At the author's institution, work at BSL—1 tends to be limited to work using non-recombinant variations of agents such as **E. coli K12** (a variety of **E. coli** so heavily adapted to artificial culture that it is no longer able to survive within a human host) or **S. cerevisiae** (common baker's/brewer's yeast). Other extremely-low-risk biological work might also be performed within a BSL—1 laboratory.

The main physical facilities required for a BSL—1 laboratory are limited to a sink for hand washing, a Biohazard sign (see Figure 8.1), and suitable facilities for biological waste disposal. Nearly any modern laboratory can be used for work at BSL—1.

Even work at BSL—1 requires attention to good laboratory practices. As in any lab, a laboratory coat, protective eyewear, and gloves should be worn. (Latex or nitrile disposable gloves are generally suitable for biological work, but when chemicals—particularly toxins—are used as part of the experiment, glove material should be chosen appropriately.) Washing hands after removing gloves is also an essential practice at any Biosafety Level. A listing of other appropriate laboratory practices can be found in [90], pp 30–33.

Figure 8.1: Biohazard sign on BSL–1 laboratory door

8.3.2 Biosafety level 2 (BSL–2)

BSL–2 is the usual containment level for the vast majority of clinical and biomedical research laboratories; most agents can be handled safely using BSL–2 practices in a BSL–2 laboratory. BSL–2 might be used for work ranging from recombinant **E. coli** experimentation to **Chlymidia sp.** studies. BSL–2 laboratories expand upon the safety measures and practices used under BSL–1.

Most BSL–2 laboratories contain **biological safety cabinets** (BSCs; see section 8.5 below), enclosed cabinets which provide protection for the laboratory occupants, the public, and also the cultures being used by constant, recirculated air filtration. (See Figure 8.2.) While BSCs sometimes are used in BSL–1 labs, their use is only to provide protection to the experiment (from cross-contamination with other cultures or from adventitious organisms in the laboratory air). The laboratory ventilation system is usually designed to maintain the laboratory under **negative pressure** relative to the outside hallways/rooms and the building surroundings. This prevents the spread of released material in the event of a spill.

BSL–2 work requires stricter controls on laboratory practices and behavior than BSL–1, but the basic concepts are the same: prevent opportunity for release of material or exposure to the investigator. See section 8.4 below for more detail.

8.3.3 Biosafety level 3 (BSL–3)

BSL–3 laboratories are considered **high-containment** facilities, and they are engineered much more carefully to prevent the spread of materials. They are far less common than BSL–2 facilities, and they are usually designed specially for the work to be conducted in them. For example, at the author's institution, the sole BSL–3 laboratory is used for studies involving **Mycobacterium tuberculosis (TB).** TB is readily spread by inhalation, so it merits a heavier level of containment than an organism such as human immunodeficiency virus (HIV). The existence of treatments for TB is one of the factors permitting it to be handled at BSL–3 instead of BSL–4, although the emergence of multiple-drug-resistant TB strains is dictating tightening of the procedures in BSL–3 laboratories handling them in order to reduce the additional researcher and community risk. In the United States, BSL–3 laboratories are annually inspected and accredited by external teams of investigators.

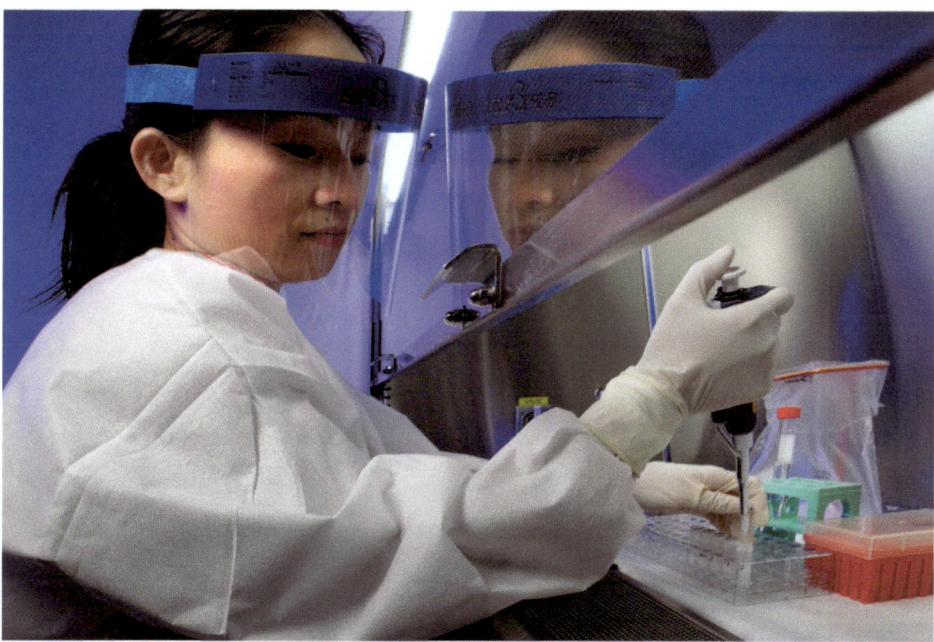

Figure 8.2: US Centers for Disease Control Biologist Shon Workman using biological safety cabinet. CDC Public Health Image Library (PHIL) #10024, photo credit James Gathany, content credits Hsi Liu, PhD/MBA; James Gathany.

BSL–3 labs vary in a variety of construction details, providing additional containment over BSL–2. For example, break-resistant windows, hands-free washing stations, and provisions to prevent escape of experimental animals are all provided. BSL–3 laboratories are usually completely sealed to prevent release of biological materials through wall penetrations, cracks between wall and floor, etc.

The heating, ventilating, and air-conditioning (HVAC) system of a BSL–3 laboratory is designed for negative pressurization and preventing recirculation of BSL–3 lab air to other parts of the building. Alarms provide warning of HVAC problems, while HEPA filter banks are used to protect the community from the contents of the lab air, when the agent in use justifies it, as with TB. Airflow is directional, into the laboratory only, and some sort of visual indicator is used to assist lab personnel in verifying appropriate directionality.

The HVAC system and the entire laboratory are designed to facilitate decontamination and sterilization; in the case of the HVAC system, seals are installed so that the ductwork can be decontaminated using formaldehyde or similar gases. Generally an autoclave is provided for decontamination of waste, although World Health Organization guidelines permit removal of infectious waste in sealed, leakproof containers for decontamination elsewhere [92].

Training and qualification is required of workers in BSL–3 laboratories. Competence is expected at a much higher level than with BSL–2 work. Researchers wear gowns, not lab coats (solid-front). Dress-in/shower-out facilities are often provided for certain organisms. (See agent descriptions in [90] for details.) Staff working in BSL–3 laboratories typically receive medical monitoring to ensure that they do not exhibit symptoms of exposure to the agent(s) in use in the lab, and so that treatment may begin immediately upon detection of such symptoms. All work with potentially infectious material is done in biological safety cabinets, not just operations considered likely to produce aerosols.

8.3.4 Biosafety level 4 (BSL–4)

The BSL–4 **maximum-containment laboratory** is used only for study of the most dangerous organisms, Risk Group 4 organisms for which there is a high researcher or community risk in the event of escape from containment and for which there is no viable treatment.

The classic modern example of such an organism is **Ebola hemorrhagic fever virus**, five species of which exist, and four of which are known to cause fatal bleeding and fever in humans. (The fifth species, **Reston virus**, has been shown to affect lower primates but not humans.) The mortality rate from Ebola virus disease is widely variable but averages 50% in a given community outbreak; mortality has approached 90% in some outbreaks. No reliable treatment is available for Ebola virus disease, although various vaccines and antivirals are in accelerated testing.

BSL–4 laboratories are distinguished by the use of **primary containment**: glove boxes in which all work with infectious material is performed (see Figure 8.3), or protective, pressurized "moon suits" supplied with clean outside air. Either the work or the researcher is fully encapsulated to prevent exposure. Personnel shower-in, shower-out (with antiseptic such as 4% Lysol®—Reckett Benckhiser—in the case of suit labs; see Figure 8.4), and change clothes completely.

BSL–4 labs are kept under a high level of security. Mechanical, electronic, or biometric locks are used to deny access to unauthorized personnel. A security-controlled airlock is used to control access to the work area.

Careful consideration is given to HVAC design: the flow of HEPA-filtered air is from the changing areas into the work areas, with all effluent being double-HEPA-filtered before discharge. Liquid effluent is sterilized before release to the environment. All other material is sterilized using a validated method (liquid or gaseous sterilant, autoclaving, etc.).

Figure 8.3: US Centers for Disease Control & Prevention BSL—4 "Cabinet Lab," c. 1978. CDC Public Health Image Library (PHIL) #7331, photo credit unknown, content credit Betty Partin.

Figure 8.4: US Centers for Disease Control & Prevention Special Pathogens Branch researcher in decontamination shower before laboratory exit. CDC Public Health Image Library (PHIL) #10729, photo credit James Gathany, content credit Dr. Scott Smith.

Emergency power supplies are available, and in many cases, redundant or back-up systems such as HVAC and effluent treatment are available.

Material is only removed from the laboratory in non-breakable, sealed containers. Decontamination of the container (and of equipment or other materials removed from the lab) is by dunking or showering in disinfectant or by fumigation with sterilant such as paraformaldehyde. With glove boxes, nothing enters or leaves without passing through an autoclave antechamber.

Because BSL–4 laboratories are so heavily sealed, special provisions for emergency communications and response are needed. Typically, specialized teams are available to perform emergency entry into and decontamination of equipment or personnel. Work alone in a BSL–4 laboratory is absolutely forbidden—a "two-man rule" applies [92].

8.4 Biological laboratory work practices

Good biological laboratory practices are a complex topic, often subject to government regulation (particularly when medical products are involved), and acquiring adequate skill in the biological laboratory requires several years of experience under competent supervision. The practices below call out a few items particularly critical to safe laboratory operations. Refer to [90, 92], and [94] for details on appropriate and safe practices, and consult with local biosafety professionals about what additional measures may be necessary for a specific organism.

8.4.1 General laboratory practices

- **Dispose of sharps** in a sharps-disposal box or (for non-disposable sharps) by placing in a hard-sided autoclavable container for decontamination. Use a blunt cannula in place of a hypodermic needle whenever possible.
- **Decontaminate all waste** containing biological material using either liquid disinfectant or autoclaving, depending on the nature of the material and the instructions of a biosafety professional. Typically, hard-surface materials and waste cultures are decontaminated by liquid disinfection, while waste bags or hard-to-wipe equipment (e.g., pipette racks) are autoclaved.
- **Transport biological materials** in sealed, leakproof containers only. Carrying covered Petri dishes from laboratory to laboratory creates an increased risk of spreading contamination or of outright breach of containment should the experimenter drop the dishes.
- **Ensure any material planned for autoclaving is compatible** with the autoclave conditions; in most laboratories, if you place non-autoclave-compatible materials in the autoclave, you will be assigned to clean up the resulting melted mess. Note that any liquid containers placed in an autoclave **must** be unsealed or contained in self-venting bottles—slightly opening or "cracking open" a bottle top may not actually permit release of pressure in the bottle. Sealed containers typically explode in the autoclave from their own internal pressure.

- **Do not allow biological material to be drawn into vacuum systems.** Culture medium is often removed by suction provided by an internal vacuum system or vacuum pump. At a minimum, most institutions require a biological filter to be installed just before the connection to the building vacuum or vacuum pump. It is recommended that a large vacuum filter flask be inserted in the line upstream of the final biological filter to collect used culture medium; the flask should be primed with disinfectant if at all possible so that collected medium is immediately disinfected. When using pathogenic or otherwise infectious material, a bubbler flask or liquid trap filled with disinfectant should also be used to protect the vacuum system. In many cases, a double-flask system is recommended: a "knockout flask" to catch aspirated material and a disinfectant bubbler to kill any aerosols that pass through the knockout flask.
- **Ensure all vacuum flasks** are either constructed of disinfectant-resistant and sterilizable plastics (e.g., polycarbonate or polypropylene) or are securely taped to avoid sprays of glass shards in the event the flask implodes. Implosion is particularly common in systems where the vacuum is especially low or where the flask is scratched or otherwise damaged. Check vacuum flasks for cracks and scratches before using, and do not use damaged equipment.

8.4.2 Personal protection

- **Wash hands before leaving the laboratory.**
- **Never leave the lab while wearing gloves.** If gloves are necessary in order to transport materials between laboratories, place the materials in a one-handed carrier and use a glove on the hand carrying the materials only. Do not operate door handles, elevator switches, etc. while wearing gloves; this may spread contamination and will also cause unnecessary (or in some cases, appropriate) concern among non-laboratory personnel such as support staff, students, and passers-by.
- **Wear appropriate eye protection.** While eye protection is not strictly necessary for most work in biological safety cabinets, it is still recommended to protect against other hazards such as imploding vacuum flasks and splashes from neighboring work or when using pathogenic material. Eye protection used in work involving hazardous or potentially-infectious biological material must be decontaminated just like other laboratory equipment; note that few models of safety glasses or goggles are autoclavable. Face protection should also be used where splashes are a risk.
- **Wear appropriate hearing protection** when using sonicators, or enclose the sonicator in a soundproof enclosure. Consult with a qualified industrial hygienist for advice on choosing appropriate hearing protection for a given application.

8.4.3 Pipetting, syringing, and other sample-transfer methods

- **Use pipetting aids**, do not pipet by mouth. You will inhale aerosols produced by the suction.
- **Use pipets with cotton plugs** to prevent contamination of pipetting aids.
- **Do not forcibly expel or "blow out" pipets.** This creates large volumes of aerosols. Use "mark to mark" pipettes in preference to blow-out type.

- **Avoid mixing solutions by repeated suctioning and blowing out of pipets**, especially with infectious material.
- **Syringes with hypodermic needles are not substitutes for pipets**—they are an invitation to a sharps injury. If syringes are absolutely necessary, use the shielded variety (see section 6.2.2 above). Note that tools are available to remove the septum from vials without creating aerosols, allowing the use of pipets for sample transfer instead of hypodermic needles.
- **Use disposable transfer loops** in favor of sterilizable ones. Although the disposable items increase waste volume, they remove the need for heat-generating sterilization equipment in the BSC. If reusable loops are necessary, use a micro furnace or electric loop sterilizer in preference to open-flame apparatus.

8.4.4 Equipment use

- **Open aerosol-generating equipment only inside a biological safety cabinet.** This includes centrifuge tubes and safety cups, blenders/stomachers, homogenizers, and sonicators, among other equipment.
- **Sonicators, stomachers, and other equipment** where mechanical energy is added to a liquid tend to build pressure in sealed vessels. Avoid the use of glass vessels for such equipment in favor of unbreakable plastics. Pressure-venting caps that filter biological materials from the released air are also available for certain applications.
- Prevent release of aerosols and cross-contamination of cultures by **holding opened tubes or bottles in a tilted or horizontal position where possible**, and by not placing bottle or flask caps on the floor of the BSC or bench top.
- **Always disinfect laboratory equipment at the end of a work session** using appropriate methods and disinfectants. Bring refrigerated or cryogenic apparatus such as cryostats to room temperature before disinfection.

8.4.5 Storage, inventory, and labeling

- **Label all stored biological materials and keep a matching inventory book or database.** Autoclave and discard any unidentified or unlabeled samples.
- **As part of check-out procedures for departing laboratory workers**, identify all samples belonging to that individual and either formally transfer them to the custody of another researcher (if they are still needed for research) or destroy them by autoclaving.
- **Periodically decontaminate not only the biological safety cabinets but also incubators, refrigerators, and freezers.** This helps remove any ancillary contamination in the laboratory and also reduces microbial load and contamination brought into the BSCs when work is performed [90].
- **Do not store materials in the liquid space of a cryogenic freezer**; they may absorb liquid nitrogen and explode upon defrosting. Most modern cryogenic freezers are designed for sample storage in the vapor space; do not overfill freezers with liquid nitrogen. Also note that viable organisms and viruses have been recovered from the liquid nitrogen phase of cryogenic freezers.

8.5 The biological safety cabinet

One of the key pieces of research safety equipment used with biological materials is the **biological safety cabinet**, or **BSC**. The BSC differs from a chemical fume hood in that it provides protection not only to the user, but also to the research samples and to the environment to which the cabinet exhaust is directed. Typical BSCs provide **personnel, product, and environmental** protection, while a typical chemical fume hood is configured for personal protection only: there is no provision made to prevent contaminants from settling into or onto research materials, nor (unless the fume hood exhaust is specially filtered or treated) does the chemical fume hood make any pretense at protecting the environment beyond air dilution of the exhaust.

There are several types of BSC, and the reader is referred to [92] for further details. The Class II Type A2 cabinet is the most common used in the modern biological laboratory. Figure 8.5 shows a typical Class II Type A2 BSC.

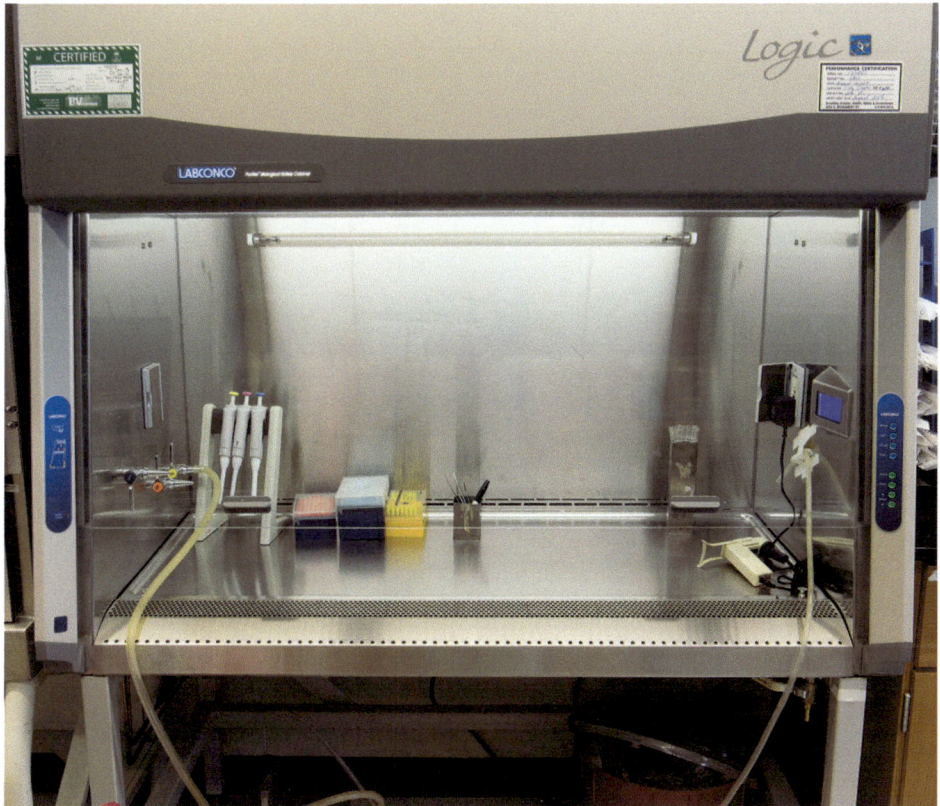

Figure 8.5: Proper setup of biological safety cabinet.

Figure 8.6 illustrates the airflow pattern inside a Class II Type A2 biological safety cabinet. Outside (laboratory) air is first drawn into the cabinet and down into a plenum through a set of grilles located at the very front of the hood; this prevents outside air from mixing directly with the air inside the cabinet. It also prevents air from inside the BSC from escaping through the front of the cabinet, providing **personnel protection** to the user and any other laboratory occupants.

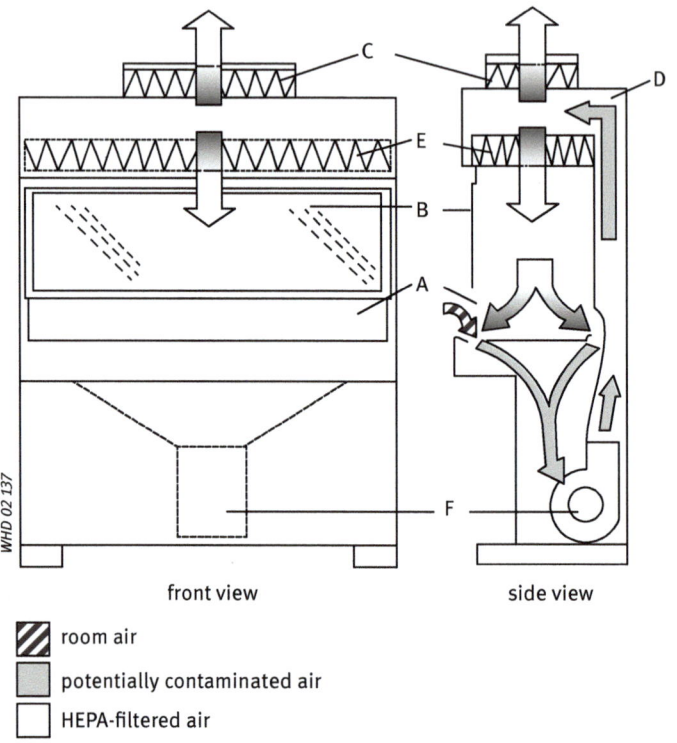

Figure 8.6: Class II Type A2 biosafety cabinet airflow detail.

The ingested air passes through a plenum at the back of the cabinet (or sometimes at the sides) to be blown down through a HEPA (high-efficiency particulate air) or ULPA (ultra-low penetration air) filter **gently** across the work surface.

The flow through the cabinet down to the work surface is non-turbulent—the air moves in separate layers (lamina), so the flow is called **laminar flow**. The laminar flow helps prevent entrainment of particulate material from inside the cabinet, which might re-deposit on research samples or accidentally escape the cabinet. This provides **product protection** to the research materials.

In a Class II Type A2 cabinet, approximately 70% of the airflow within the cabinet is drawn through the grilles at the front of the work area and recycled; in some cabinets, the remaining 30% is discharged. The discharged material passes through another HEPA or ULPA filter before discharge to an exhaust duct (or into the laboratory, depending upon the cabinet design). This final filtration provides **environmental protection**, removing any entrained hazardous biological material before it escapes the cabinet.

8.5.1 A BSC is not a chemical fume hood

A BSC is **not a chemical fume hood**. Because of the air recycling scheme used in biological safety cabinets, it is unwise to use substantial amounts of toxic or flammable materials in a BSC—the air recycling can cause buildup of hazardous toxic or flammable quantities of material rather rapidly, especially for volatile compounds. While the author has never located a substantiated incident of a BSC exploding from excessive use of flammables, the possibility exists. When sterilizing a BSC with flammable sterilants (e.g., 70% ethanol), spray or pour sterilant on sterile wipes well **outside** the cabinet before use.

Use of hazardous chemicals should be kept to an absolute minimum, and chemicals should **never** be stored in a BSC. If it is necessary to use hazardous chemicals in a BSC (e.g., in order to dose cultures with toxins), work as closely as possible to the rear of the cabinet, near the baffles. This will lower the amount of chemical that leaks from the front of the cabinet. **Note that this method of protection works only with BSCs vented directly to the outdoors.** If the BSC is vented to the room, **do not use the cabinet with toxic or flammable chemicals under any circumstances without consulting a qualified industrial hygienist or safety professional for advice.**

Class II Type B2 biological safety cabinets are 100% exhausted and can be (if designated by the manufacturer) used for flammable and toxic materials, provided they exhaust to an appropriate location outdoors rather than into the lab.

8.5.2 The "laminar flow hood" or "clean air hood" is not a BSC

As noted above, biological safety cabinets are sometimes called **laminar flow hoods** for the non-turbulent flow of air established within them to prevent intrusion of contaminated air into the workspace. Unfortunately, the term does **not** apply only to BSCs—there are other types of hood which require similar contaminant-free work areas and use technology similar to the BSC to establish a laminar flow of clean, filtered air across the work surface.

Electronics assembly and other similar operations are a common example. Figure 8.7 shows a horizontal-flow "clean air workstation" used for electronics work. This cabinet provides a constant flow of filtered air across the work surface, mainly to exclude particulate contamination. It might be used, for example, for cleaning the sensor of a digital camera, to prevent settling of additional airborne particulates on the sensor surface (which would obscure the images made by the camera).

The clean air workstation has a simpler air flow path than the biological safety cabinet. Air is drawn in through the top or rear and passed through a HEPA or ULPA filter for cleaning, but it is then directed down upon or across the work surface, normally exiting through the front of the cabinet. This means that the cabinet air is being blown directly into the face of the user. While the flow is laminar, the filtered air leaves the cabinet through the front rather than through a filtered exhaust port like a BSC.

For this reason, clean air cabinets provide no personnel or environmental protection. **Never use a "clean air hood" or similar device for research operations with hazardous biological or chemical materials.**

Figure 8.7: Clean air workstation.

8.5.3 Using a BSC

A biological safety cabinet is similar to a chemical fume hood (see section 7.3 above) in that it relies on a defined airflow to perform its job to maximum efficiency. Consequently, airflow disturbances should be avoided to the maximum extent possible. Inserting/removing items, including hands, drags contamination from the laboratory air into the BSC, and obstructing any of the air-control ports (such as the register at the front of the cabinet or the baffles at the rear) retards removal of those contaminants. The guidelines below are written with this in mind.

Note that many of the practices below work to protect not only the BSC operator and the other occupants of the laboratory but also to protect the work itself from inadvertent contamination. Cutting corners and ignoring proper work practices will lead to contaminated samples and non-reproducible results. Most advice below is summarized from [92] and [90].

8.5.3.1 Loading and setup

- **Stage all needed materials inside the BSC when starting work.** This minimizes round-trips of the hands in and out of the cabinet. Decontaminate the surface of all materials with 70% ethanol or other disinfectant before insertion. Allow several minutes for the airflow in the BSC to remove airborne contamination that entered with the supplies and other materials.
- **Do not stage or store materials (extra gloves, wipes, clean culture flasks) beyond that necessary for the work to be performed inside the BSC.** The BSC should contain only materials planned for immediate use during the current work session.
- **Stage supplies on the left-hand side of the cabinet and locate materials disposal (waste culture medium containers, autoclave bags, etc.) on the right.** Use the center of the cabinet for the actual research work. For extra protection when using pathogens or for extra-sensitive work, place a disinfectant-soaked absorbent pad at the center to kill any incidental splashes and droplets. Otherwise, place one or more dry absorbent pads on the work surface to absorb splashes and prevent them rebounding into inhalable aerosols. Many BSC users will use multiple layers of absorbent pads that have non-absorbent bases—in the event one pad becomes contaminated or overly dirty, it can be removed quickly and work can continue immediately.
- **Place material as far toward the back of the cabinet as possible without blocking the rear air grilles or baffles**—the possibility of recirculation or escape of material is minimized. Aerosol-generating equipment such as centrifuges or sonicators should also be located in the rear.
- **If the BSC is equipped with a switch-operated air-circulation system, turn it on** and wait 5 minutes for the cabinet to filter existing particulates from the cabinet interior.
- **If an ultraviolet lamp is provided in the BSC, turn it off before working in the cabinet.** Most modern cabinets have interlocks that prevent operation of the UV lamp when the sash is open, but older cabinets may have manual switches. (See section 9.2.1 below for details on UV light hazards.) Note that the effectiveness of UV lamps depends strongly on a variety of factors, including the age and cleanliness of the bulb, location of shadows within the cabinet, UV exposure time, and the sensitivity of the organisms in use to UV light. For this reason, neither the World Health Organization nor the US Centers for Disease Control and Prevention recommend the use of UV lamps in biological safety cabinets [90, 92], and the latter reference goes so far as to call them unnecessary and not recommended.

8.5.3.2 Working in a BSC

- **Upon inserting hands into the BSC, pause for approximately 1 minute** to allow the airflow to sweep contamination from hands and arms. The act of breaking the air-barrier at the front of the BSC brings with it a certain amount of non-sterile laboratory air; the pause allows removal of any particulates entrained in this introduced air.
- **If the BSC has a movable sash, keep the sash at the recommended position.** Many newer BSCs are equipped with alarms that will sound if the sash is opened too far. Close the sash when the BSC is not in use, both to avoid spread of contamination and to save energy (since expensively conditioned laboratory air is constantly being drawn out of the lab through the BSC if it is a type vented to the outdoors).
- **Do not allow materials or arms to rest upon the grille at the front of the BSC**, as this drastically affects airflow in the cabinet. See Figure 8.8.
- **Do not walk close to the front of a BSC**, especially if it is in use; the wake cast by the body will create a negative pressure wave, pulling air out of the BSC (and into the operator's breathing zone). Similarly, avoid rapid opening and closing of doors while BSC fronts are open.

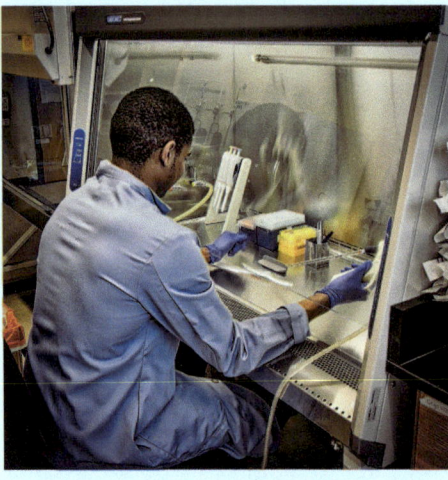

Figure 8.8: Researcher Quinton Smith poses to illustrate improper use of BSC.

- **Work from left to right**, with the left-hand side of the cabinet maintained as the "clean" side and the right as the "dirty" side. Mixing clean and dirty supplies leads quickly to cross-contamination of samples; also, dirty supplies may be mistaken for clean, removed from the cabinet at the end of work, and re-used later.
- **Do not remove material from the BSC until work is complete if at all possible.** Place biohazard waste or used pipettes in intermediate containers within the BSC while working rather than discarding them in exterior bags, boxes, or other collectors. Using an external trash container leads to repeated removal and re-insertion of hands into the cabinet, increasing the amount of external, un-cleaned air in the cabinet.
- **Dispose pipettes in horizontal, disinfectant-filled trays or other containers**—using a vertical pipette disposal can leads to excessive arm movement and spreads contamination [90].
- **Avoid the use of open flames:** gas burners should be equipped with foot-operated self-closing gas valves that shut off should the researcher lift his or her foot. Alcohol burners have been the source of many fires in BSCs and should be avoided at all costs. Note that electric loop

sterilizers and other similar apparatus are available to perform the tasks previously requiring alcohol or Bunsen burners. Open flames also cause turbulence inside the laminar-flow environment of the BSC, degrading its performance.

8.5.3.3 BSC maintenance
- **Upon completing work in a BSC, fully seal all cultures, biohazard bags, etc. before removal.**
- **Disinfect all materials** using 70% ethanol, bleach solution, or other appropriate disinfectant before they leave the BSC. This includes "clean" supplies to be returned to storage and the exterior of waste bags.
- Once the BSC is empty, **disinfect the BSC with suitable disinfectant**. When using 70% ethanol, spray laboratory wipes with disinfectant **outside the cabinet** to avoid creating aerosols of flammable materials inside the cabinet. Note also that chlorine bleach is highly corrosive toward the stainless steel from which most BSCs are made. Bleach should not be allowed to sit wet on BSC surfaces longer than recommended by a biosafety professional for satisfactory disinfection; rinse the cabinet by wiping it down with sterile water after using bleach.
- **BSCs require annual certification**, and depending on use, may require annual or more frequent replacement of HEPA filters. Certification and HEPA filter change should be conducted only by qualified personnel.

8.6 Case studies

8.6.1 Biology experiment

The biology of learning, a branch of behavioral biology, concerns itself with studying how neurological systems and circuits self-organize and respond to various stimuli. Sometimes the learning is beneficial, such as when an animal learns how to find food, but other times learning mechanisms can become pathologically compromised. Drug addiction is the classic example. A laboratory studies drug addiction by examining how cues gain biological importance, whether that cue is a bell in a classic Pavlovian experiment causing a dog to salivate in anticipation of food or the cue is the sight of a liquor bottle causing relapse in a recovering alcoholic. Certain parts of the brain, such as the ventral tegmental area (VTA), play a major role in the biology of addiction; neural "projections" from the VTA to other brain structures such as the amygdala or the lateral hypothalamus seem to be involved deeply in learning processes [95].

Modern practice [96] permits individual neuronal circuits to be studied in great detail through optogenetic methods. In these techniques, mice or rats are injected in specific locations with specialized viruses that contain genes coding for light-sensitive proteins; the mice have been genetically-engineered so that they will express the viral payload, creating the light-sensitive protein in quantity. These proteins typically affect the action of ion channels, modulating the response of the neuron. The target neurons can then be forced to fire (or not to fire) using tuned laser light applied to the brain tissue **in vivo**. Electrodes implanted in other locations allow study of the effect activation of the target region has on other brain structures, and direct behavioral observation can be made in apparatus that provides the mouse or rat with a reward (such as cocaine) under specific conditions. (See Figure 8.9.) Thus, specific studies can be made of such things as dopamine-dependent (dopaminergic) neurons in the VTA region. Interestingly, it has been found that dopaminergic activity, which is heavily involved in the biology of addiction, depends not directly on the reward stimulus (e.g., the drug) but on the **difference** between the reward received and that which was anticipated [97].

Figure 8.9: Apparatus for mouse addiction testing.

Several biological hazards exist in this type of research. First and foremost is the use of engineered viruses—many viral vectors (e.g., **vibrio** sp.) are infectious, and even those which are not may have unpredictable effects once engineered. Animal models such as mice and rats, while normally maintained in a sterile environment, may carry infection; they are capable of biting and causing injury or transmitting various diseases. Since microscopic and immunohistochemical studies are part of the research, the researchers may be exposed to blood and tissue, which again may carry infectious material. The risk of infection is raised because of the above-mentioned potential for animal bites and because of the use of scalpels and hypodermic needles—a bite, needlestick, or scalpel cut bypasses the protection of the skin and provides any infectious matter direct access to the bloodstream.

Biological hazards are controlled using a combination of work practices and experimental design. First and foremost, the viral vectors used are **not capable of replication**; this is another example of **inherently safer design** (see section 15.3 below). The experiment is made safer by design conditions such that certain risks do not exist or are drastically reduced. In addition, surgeries are conducted using standard personal protective equipment for BSL—2 conditions: lab coat or gown, gloves, mask, and safety goggles (as opposed to safety glasses because of concerns about aerosolization of material). Some organizations even conduct work such as this at BSL—1. Blood & tissue are decontaminated by a 15 minute soak in 10% sodium hypochlorite and bagged for incineration. Hypochlorite is also used for daily and weekly deep cleaning of all apparatus. Pathology work such as cutting with a cryostat is performed only on fixed tissue to avoid aerosolization of potentially-infectious materials.

8.6.2 Civil/environmental engineering experiment

When we send water down the drain, most of us think it "goes away," disappearing in some magical manner to somehow be recycled by nature's water cycle into fresh water from the tap. The reality of the process is starkly different. Wastewater in most developed countries flows through sewer systems to wastewater treatment plants like Baltimore's Back River Treatment Plant in Maryland (US), where complex biological processes are employed to render the wastewater clean enough for discharge into the natural environment.

Not surprisingly, "primary wastewater," the inflow into the treatment plant, is highly nutrient-rich, and once sterilized, these nutrients, particularly nitrogen and phosphorus, become available to feed other microorganisms. If released directly into local aquatic ecosystems, the additional nutrient load would cause large-scale damage to the ecosystem, drastically disrupting the balance of microorganisms and leading to phenomena such as algal blooms and "red tides."

One of the primary purposes of wastewater treatment is to reduce the nutrient content of discharged treated water. Algae of various types play a critical role in removing these nutrients from the flow, allowing them to be separated as biomass, either in sludge or separately. Research by Bohutskyi et al. [98] studied the growth of eighteen varieties of microalgae on unsterilized primary wastewater and on a nutrient-rich intermediate wastewater product called Anaerobic Digester Centrate (ADC). The purpose of the study was to identify microalgae suitable for both nutrient removal and accumulation of biomass in the wastewater treatment plant: nutrient removal for the reasons given above and biomass production as a route to production of biofuels such as biodiesel produced from algal oils.

The study used samples of untreated wastewater and ADC obtained from the Back River Treatment Plant. Naturally, these samples, particularly the untreated wastewater, contain the full range of materials that go into sewers, and many of those materials are biohazardous. Concentrations of **E. coli**, for example, ranged from 10^5–10^7 cells/L, and other bacteria such as **Enterococcus** and **Pseudomonas aeruginosa**, both of which are pathogenic, were present in large quantity.

In this case, normal microbiological practices followed to prevent cross-contamination of samples sufficed to ensure researcher safety. Work involving unsterilized wastewater was conducted in biological safety cabinets using BSL–2 work practices in a BSL–2 laboratory. Figure 8.10 illustrates the algal growth trials taking place in a laboratory BSC. Note the secondary containment trays provided to catch spills or leaks. Secondary containment should always be provided for hazardous liquids, biological or otherwise; the containment should be large enough to contain the largest expected spill, in this case one of the Erlenmeyer flasks being used as bioreactors.

Figure 8.10: Algal growth trials in biological safety cabinet.

8.7 Exercises: biological hazards

1. Using **Biosafety in Microbiological and Biomedical Laboratories** [90] as a reference, select those practices and laboratory design features suitable for **BSL–2** operation.
    a. Facilities for collection and disposal of biological waste
    b. HEPA-filtered ventilation system
    c. Air-supplied, sealed "moon suit"
    d. Class II Biological Safety Cabinet
    e. Glove box (Class III BSC)
    f. Disposable nitrile gloves
    g. Lab coat or surgical gown

h. Negative-pressure ventilation system
   i. Biohazard signage
   j. Access to autoclave
8. Again using the "BMBL," identify the appropriate Biosafety Level for the work described.
   a. Brewing process research using **Saccharomyces cerevisiae** (brewer's yeast)
   b. Molecular biology manipulations using **E. coli** (a strain which produces Shiga toxin/Verocytotoxin, which causes severe diarrhea)
   c. Large quantity production of **Bordetella pertussis** (the causitive agent of whooping cough/pertussis).
   d. Laboratory cellular studies of LGV serovars of **Chlamydia trachomatis.**
   e. Cell culture of Severe Acute Respiratory Syndrome (SARS) coronavirus.

9 Radiation hazards

In this chapter, we consider **radiation hazards**, that is, hazards from energy transmitted through space, whether that energy is electromagnetic in nature or kinetic energy from a subatomic particle or an atomic nucleus. Radiation hazards are divided into two basic categories, depending on their mechanism of action.

9.1 Ionizing radiation

Ionizing radiation is radiation of sufficient energy to ionize matter. **In vivo,** ionizing radiation can act directly on DNA, but the more common mechanism of action is the creation of cascades of water radicals and radical ions such as ·OH and ·(−OH), which themselves act upon DNA, proteins, and other biochemical constituents of cells to create overall havoc.

9.1.1 Types of ionizing radiation

Ionizing radiation comes in two varieties:
- **Particulate radiation** is radiation from radioisotopes, nuclear processes, accelerated electron or proton beams, and particulate cosmic rays such as high-energy protons or antiprotons, etc.
- **Ionizing electromagnetic radiation** is high-frequency/short-wavelength electromagnetic waves, generally in the X-ray region or higher frequency. Very high energy ultraviolet light (UV-C) can also be ionizing.

Each type of ionizing radiation has different properties and can require variations in control measures. For example, while lead bricks are excellent radiation shielding for gamma rays, thin sheets of lead are a poor choice for protection from electron beams due to secondary X-radiation called **Bremsstrahlung.**

9.1.2 Sources of hazard from ionizing radiation

9.1.2.1 Particulate radiation
Particulate radiation is encountered in two major ways: as products of radioactive decay from radioisotopes or as organized beams of particles in particle accelerators, from cyclotrons and synchrotrons to the tiny electron accelerator in a cathode-ray tube.

Alpha particles

Alpha particles are bare helium nuclei: two protons bound to two neutrons. (Chemically, they are $^4\text{He}^{2+}$ ions.) Typical alpha particles from radioactive decay move at approximately 5% of lightspeed, which combined with their mass gives them a high kinetic energy. They interact strongly with other matter, which is a mixed blessing: a few inches of air, a sheet of paper, or a laboratory coat suffice to shield the researcher from the average alpha radiation source. Nevertheless, the high energy means that if an alpha source such as ^{239}Pu is inhaled or ingested, most or all of the particle's energy is likely to be deposited in the lungs, stomach, or any organs which concentrate the element. This makes alpha-generating materials particularly dangerous to handle in experiments where they may become airborne. Certain modes of fission can produce much higher energy alpha particles, and alphas used in cyclotrons or other particle accelerators can reach high energies.

Beta particles

Beta particles are electrons or positrons emitted from radioactive sources such as tritium (^3H) or phosphorus-32 (^{32}P); electrons may also be produced by special-purpose electrical equipment ("electron guns") such as those used for scanning or transmittance electron microscopy (SEM/TEM). Beta particles generally originate in the nucleus rather than being ejected from an electron shell. Energy levels are typically moderate (although they vary widely with the source), and for common sources such as tritium, may be shielded by a centimeter or so of polymethylmethacrylate (PMMA; Plexiglas®—Arkema; Perspex®—ICI; and a host of other trade names); some high-energy sources such as ^{90}Sr may require additional shielding and handling practices.

Neutron radiation

Neutron radiation is exclusively generated by nuclear decay and fission processes, whether those processes are natural, such as in uranium ore, or artificial, such as the enhanced neutron flux in a nuclear fission or fusion reactor. Neutron radiation can be highly penetrating and difficult to shield against.

Proton and heavy-nucleus radiation

Large amounts of proton radiation and "heavy-ion" radiation (i.e., heavier than ^4He) are unusual in laboratories outside particle physics centers and certain health care facilities. Measurement of, handling of, and shielding for proton and heavy-nucleus radiation is highly specialized; those working with it typically receive extensive training (which is beyond the scope of this book). Proton and heavy-ion radiation can be quite energetic: at the energies typically used in cancer treatment, concrete shields of 1 m or more are required in the treatment rooms [99].

Proton radiation on Earth mostly derives from the solar wind (the vast majority of which is diverted by the Earth's magnetic field, and from cosmic rays. Such exposures are a significant concern for space travel outside the Earth's heliopause, and research into portable shielding has been conducted for decades. High-energy cosmic rays (mostly protons and antiprotons) continually bombard the Earth. Energies are high enough for many cosmic rays to penetrate the Earth's magnetic field, and so proton radiation makes up a portion of the natural background radiation to which humans are constantly exposed.

9.1.2.2 Electromagnetic radiation

Electromagnetic radiation is ubiquitous in the laboratory, starting with the lighting fixtures. Lamps, however, are designed to emit incoherent visible light, rather than ionizing radiation. X- and gamma- (γ-) rays, on the other hand, are regularly encountered in biomedical and physics laboratories.

X-rays and gamma-rays

X- and gamma-rays are defined colloquially in terms of energy: X-radiation is lower-energy ("softer") than gamma-radiation. Technically, though, X-radiation is any electromagnetic radiation deriving from an electronic process, while gamma-rays are produced by nuclear decay processes. There is no exact range for either. Gamma rays tend to be in excess of 100 keV (10^{19} Hz), while X-rays are usually lower-energy, 100 eV to 100 keV. Those X-rays toward the higher-energy end of the range are referred to as "hard" X-rays.

Generation of EM radiation by particulate sources

Most gamma-rays are produced by radioactive decay processes, although some astrophysical phenomena (such as pulsars) and even strong thunderstorms are known to emit them. While there are many astrophysical sources (e.g., the Sun), X-radiation reaching the Earth's surface is typically artificially produced through one of two methods:

1. **Characteristic X-rays** can be produced by electron bombardment of heavy metal targets in a vacuum tube. In this process, an electron of sufficient energy can knock out an inner-shell electron of the metal, leading to emission of an X-ray as an outer-shell electron falls to fill the hole created. Characteristic X-rays may also be created by nuclear decay processes such as K-electron capture.
2. **Bremsstrahlung** (or "braking radiation") is X-radiation produced by the motion of electrons as they pass near heavy metal nuclei. Due to the strong electric field, the electron is forced to make a sharp change in direction, which leads to emission of a photon carrying the energy lost. In addition to braking radiation produced in X-ray tubes, special purpose accelerators called **synchrotrons**

use this effect to generate very strong X-ray beams for scientific and medical research.

9.1.3 Control of ionizing radiation

9.1.3.1 Avoidance of exposure
The best way to reduce exposure to a hazard is to avoid exposure in the first place. Not being where the hazard is, or better yet, never creating the hazard in the first place are excellent techniques to accomplish this.

Radioisotope use in biological research, a previous heavy user of common isotopes, is dropping with the rise of more sensitive alternatives to radioassay techniques. With the closure of certain production reactors due to aging, exotic isotopes have become more difficult to obtain; shortages of ^{99}Mo, the precursor to the medically-essential ^{99m}Tc (the "m" denotes that the nucleus is in a **metastable** high-energy state), have thrown medical diagnostics into disarray. This has the effect of reducing the number of scenarios through which a research laboratory worker may be exposed to radioactivity.

9.1.3.2 The three pillars of health physics
Health physicists employ the ALARA principle—As Low as Reasonably Achievable—to reduce radiation exposure using three basic methods:
- **Time.** The longer a person is exposed to a radiation source, the larger the received dose will be. Practicing radioisotope procedures with "cold" (nonradioactive) chemicals to reduce the time necessary to complete an experiment is an example of using time to control exposure.
- **Distance.** Radiation intensity falls off with the square of the distance from a point source; therefore, the further the researcher is from the source, the lower the received dose will be. Some laboratories employing very strong radiation sources will use special remote-manipulation apparatus ("waldos") to work at significant distances or behind thick shielding (see below). (See Figure 9.1.)
- **Shielding.** Many materials absorb ionizing radiation, and interposing such materials between the radioactive source and the user naturally reduces the received dose. Materials must be selected carefully for the use and the radiation to be blocked: for example, lead is a poor shielding for neutrons, as neutron capture produces unstable elements such as ^{207m}Pb, leading to secondary radiation hazards as the metastable nucleus decays and emits a γ-ray. (In Figure 9.1, the glass between the investigator and the operating compartment is thick and likely heavily-leaded to shield from gamma radiation.) Using heavy elements to shield against charged particle radiation is imprudent as it leads to secondary X-radiation produced by **Bremsstrahlung.** This is why hydrogen-rich materials such as plastic are generally used as shielding for β-radiation.

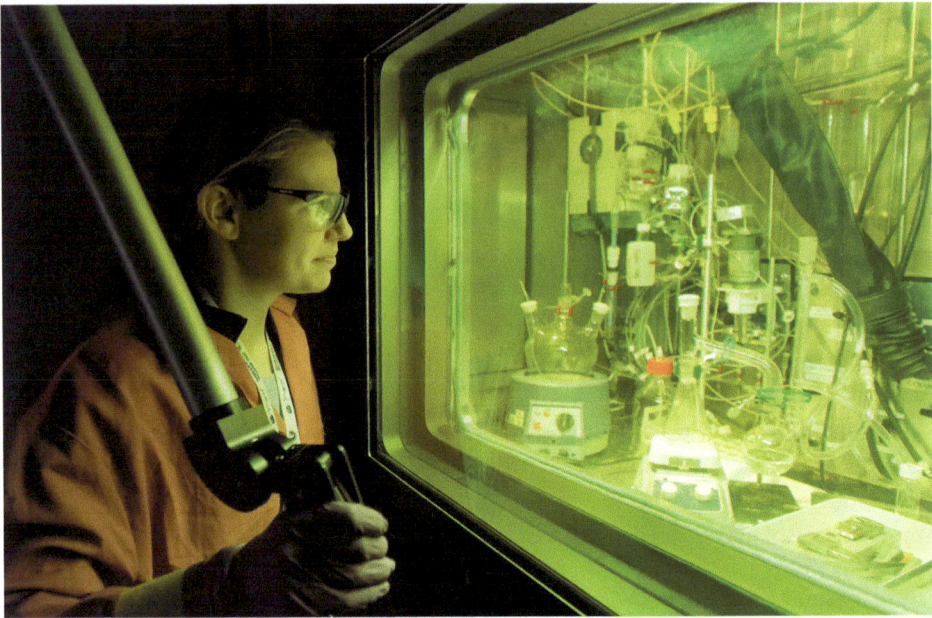

Figure 9.1: Radiochemist Dr. Amanda Youker works on techniques for ^{99m}Tc production using remote manipulators (or "waldos"). Photo courtesy Argonne National Laboratory.

9.1.3.3 Exposure limits

Establishing exposure limits is an essential first step to controlling radiation, since it provides a goal for which to aim and a criterion for judging success or failure.

Determining a universally-"safe" level of radiation has eluded health physicists since the discovery of ionizing radiation. Higher amounts, such as the 21 Sv or so received by Louis Slotin in 1946, are obviously lethal, as Slotin died unpleasantly within days. Lower levels received over time are also clearly harmful, as evidenced by the high rate of ^{222}Rn-induced lung cancers seen among uranium miners in the 1930s [100].

Low doses and dose rates of radiation are not proven to cause cancer [101]. Some health physicists (radiation-control specialists) and radiation epidemiologists argue that risk from ionizing radiation does not scale linearly; instead, low doses are theorized to induce hormesis (a favorable biological response) due to stimulation of the body's efficient DNA repair mechanisms. This hypothesis is not widely accepted, particularly by regulatory bodies and the public, who tend to hold to the "no safe dose" criterion for radiation exposure.

Such a criterion implies that all exposure to ionizing radiation should be avoided, and this philosophy leads health physicists to attempt to reduce exposure as much as possible. As mentioned above, this **As Low As Reasonably Achievable** criterion is the central tenet of radiation exposure control. The actual exposure limit for ionizing radiation varies by type of radiation (alpha particles vs. γ-rays, for example), specific energy of the particles or photons (which affects the body's absorption of the radiation and the potential for ionization), part of the body irradiated (e.g., the cornea is particularly susceptible to radiation-induced cataracts), and the population to which the exposed person belongs (e.g., pregnant women are permitted far less exposure than a normal radiation worker due to the radiation sensitivity of a developing fetus).

In order to estimate the exposure received by a radiation worker, a variety of personal monitoring devices are used, the most well-known of which is the **film badge**. As the name implies, a film badge is a flat packet worn on the lapel or shirt containing film that darkens in response to ionizing radiation exposure. Film badges, once developed, generally provide an estimate of whole-body exposure. Ring-style badges are sometimes used to measure specific exposure of the hands, such as when an experiment requires the researcher to reach behind shielding and manipulate samples in close proximity to a radiation source. Film badges have mostly given way to more sophisticated techniques such as thermoluminescent detection, except in applications where exposures are low and the application is cost-sensitive (film badges are inexpensive). The term "film badge" persists colloquially for any personal radiation-monitoring device.

9.2 Non-ionizing radiation

Not surprisingly, **non-ionizing radiation** is radiation having insufficient energy to ionize matter. Non-ionizing radiation varies across the electromagnetic spectrum from ultraviolet down to long-wavelength radiofrequency radiation.

9.2.1 Ultraviolet radiation

One common laboratory form of non-ionizing radiation is **ultraviolet radiation (UV)**, which derives from intermolecular electronic processes as opposed to actual ionization of the molecule. UV falls into one of three wavelength ranges:
- **UV-A** radiation ranges from 315–400 nm;
- **UV-B** radiation varies from 280–315 nm; and
- **UV-C** radiation covers the range from 100–280 nm.

An older system of "near," "middle," "far," "and "extreme" is also in use. (Extreme UV, ranging from 10–100 nm, is considered ionizing radiation because of its capability to interact with outer valence electrons of atoms.) See ISO 21348 [102] for further detail.

Common laboratory sources of UV radiation include ultraviolet lamps (including "black lights"), handheld UV sources such as those used for visualizing chromatography plates, UV lasers and LEDs, UV sterilization lights in biological safety cabinets, UV transilluminators used for reading DNA electrophoresis gels, UV lithography machines, and in specialized laboratories, plasmas and synchrotron sources. Welding in instrument construction shops is a source of particularly strong, broad-spectrum UV radiation.

Control of UV light is best accomplished through a combination of engineering controls and personal protective equipment. UV hazards include skin damage ("sunburn," photosensitization and photoactivation of certain pharmaceuticals) and eye damage (photokeratitis, cataracts), and other injuries or maiming, depending on the wavelength of the UV radiation and the amount of exposure. As always, use engineering controls to reduce the potential for exposure to a minimum—do not use improperly-designed or -maintained equipment that permits significant amounts of UV to "leak" from the apparatus.

Always use appropriate eye protection with UV sources. Although most eyeglasses, including standard safety glasses, do provide some protection from UV, exposure to reflected radiation from the sides (or reflected from the inner surface of the lens) can be significant, particularly with high irradiance or long-term exposure. Wear **goggles** made of inherently UV-blocking materials (such as polycarbonate) or treated with UV absorbing coatings instead. (In high irradiance or long-term exposures, always request an evaluation from a radiation safety professional—even polycarbonate may not have a high enough UV blocking capability in many applications.) Note in particular that standard polycarbonate safety goggles often do not block highly-energetic UV-C radiation effectively, and this should be taken into account when working with UV-C sources or wide-spectrum UV lamps.

If there is a danger of skin exposure, cover all exposed skin with UV-opaque clothing. Sunscreen may also be used, but again, such use should always be checked with a radiation safety professional in order to ensure that protection is adequate [103].

9.2.2 Infrared radiation

Infrared radiation (IR) is electromagnetic radiation in a wavelength range just longer than visible light; an "IR-A," "IR-B," and "IR-C" system is in use by some researchers and applications engineers [102]. IR is mostly hazardous for its heating properties, whether the IR equipment is intended to generate heat or whether it serves some other purpose. In the laboratory, IR lamps are used for operations such as curing certain polymers and heating samples (e.g., glass working); IR lasers or other strong sources may cause corneal or retinal damage. (See section 9.2.4 below.) Because of the radiative heat transfer involved in IR sources, fire is also a possibility (sometimes a significant one, such as when white-hot ceramic samples are removed from furnaces and set too near combustible materials).

As with UV light, IR light protection should be accomplished by engineering controls as much as practical. Filters, noncombustible curtains, shields, etc. may be used to block the radiation completely. If exposure is unavoidable, consult a radiation safety professional, who can determine if personal protective equipment is needed and identify appropriate types.

9.2.3 Radiofrequency (RF) radiation

Frequencies of electromagnetic radiation from about 3 kHz and above are termed **radiofrequency radiation (RF).** While it has been hypothesized that long-term exposure to RF, particularly at frequencies commonly used for telecommunications such as mobile phones, causes brain tumors, this supposition is not fully supported. RF electromagnetic fields are classified **2B-Possibly carcinogenic to humans** by the World Health Organization based on limited evidence of gliomas (a type of brain tumor) in animals and inadequate evidence of other tumorigenesis [104]. While this is not a strong classification, prudence would dictate attempting to minimize exposure to RF radiation.

As the body is mostly water and electrolytes, it makes an excellent antenna for radiofrequency radiation having wavelengths similar to the sizes of body parts (e.g., 10 cm **Extremely High Frequency—EHF**—radiation is approximately the dimension of a human head). Strong RF exposures in those frequency ranges lead to high specific absorption rates of RF energy, creating heating in the affected tissues. Eyes and testes are particularly vulnerable to RF absorption because of limited blood flow to remove heat.

Many countries publish legal standards on exposures to radiofrequency electromagnetic radiation—see, for example, [105]. Typically, the frequency range corresponding to the lowest allowable exposures is approximately 10–100 MHz; note that exposure limits for magnetic fields are typically an order of magnitude lower than for electric fields.

Shielding against RF electromagnetic fields is extremely difficult. Short of the construction of a Faraday cage around the source, it is nearly impossible to completely prevent leakage of RF radiation from an apparatus. Protective suits made of conductive fabrics are available (and are sometimes used by mobile phone tower maintenance personnel) but are also less than fully effective. This leaves **time** and **distance** as the principal methods for protecting against RF electromagnetic fields.

Exposure time may be manipulated by turning off the RF source or reducing power while personnel will be near. (Mobile phone towers, for example, are always depowered and "locked out" before service.) As exposure limits are expressed as time-weighted averages, constructing the experiment to minimize the time researchers will be exposed is also effective.

Effective field strengths (for both electric and magnetic fields) may be estimated using the methods described in [106]. Appropriate **separation distances** between researcher and apparatus can then be calculated from field estimates.

Particular attention should be paid to electromagnetic fields from directional antennae and other similar apparatus! While leakage from microwave ovens, etc. (at least those in good working condition) has been minimized by careful design over the years since microwave technology was commercialized, radiation from parabolic microwave antennae is capable of causing serious thermal damage to humans and animals from specific absorption of microwaves. The first microwave oven was designed by a researcher who noticed that his lunch became hot and was cooked when left close to an energized directional microwave antenna [107]. The same heating can happen to a human head. Distance is not a significant protective factor for directional RF sources.

9.2.4 Laser light sources

By far the most common source of hazardous non-ionizing radiation in the laboratory is the **laser**. Lasers (the name stands for **Light Amplification by Stimulated Emission of Radiation**) emit a beam of coherent (meaning that all photons are in-phase with one another), monochromatic (a single wavelength, although exceptions emitting on several discrete wavelengths are known) light in a single direction, at a high power density.

Lasers come in a bewildering variety of types and with many different wavelengths. The table below lists a few common in laboratory settings.

Table 9.1: Example lasers and wavelengths produced

Laser type	Wavelength (nm)
Excimer (ArF)	193
HeNe	543, 632
Nd:YAG	266, 532, 1064
Argon	457, 476, 488, 514
GaAs	905
CO_2	10 600

Table adapted from [43].

Most common is the **laser diode**, which operates similarly to a light-emitting diode (LED), although laser diodes consume much greater currents and emit much more light. Some lasers emit light continuously, while others operate in a **pulsed mode**; this complicates the definition of the laser's power. Beam wavelengths vary from the UV region (and occasionally beyond) to the far infrared; similar devices emitting microwaves are called **masers**.

9.2.4.1 Laser classifications
In order to more easily identify the hazardous characteristics of particular lasers, every laser (except a laser newly-constructed in the laboratory, perhaps), receives a classification under laser safety standards [43]
- **Class 1** lasers are not capable of causing harm when used normally. A **"Class 1M"** subclass is given to lasers that might pose a threat when viewed through magnifying optics (e.g., the objective of a microscope). Enclosed or embedded systems are often rated Class 1, regardless of the power of the laser inside, provided that the enclosure is not opened.
- **Class 2** lasers are technically capable of causing injury, but the eye is protected by the **aversion response,** that is, the automatic tendency of the eye to shut when exposed to bright light. A **"Class 2M"** subclass is defined for the same reasons as Class 1M.
- **Class 3** lasers are capable of causing eye injury; two subclasses exist. **Class 3R** lasers are marginally unsafe at <5 mW—from a practical standpoint, concentration of the beam by magnifying optics would be necessary to cause harm. **Class 3B** lasers (5–500 mW) are hazardous to the eye by direct viewing (see section 9.2.4.2 below) and by specular reflection. Lasers emitting invisible wavelengths are Class 3B at any power below 500 mW.

> Most "laser pointers" such as those used for pointing at slides or screens during lectures are of Class 2 or 3R. In the United States, only handheld lasers of Class 3R and below may be sold as "laser pointers." Nevertheless, "handheld lasers" and "astronomical pointers" having substantially higher power (up to the range of several watts) are also sold, and customs inspectors cannot intercept every overpowered "laser pointer" crossing the border. Researchers at the US National Institute of Standards and Technology noted recently [108] that laser pointers exhibited very poor quality control and varied substantially in power and wavelength emissions from their label specifications. Laser classification stickers can be quite small—the author once purchased a "laser pointer" and used it for lecturing for a year before reading the tiny print on the label and discovering that it was actually a 15 mW Class 3B laser device.

- **Class 4** lasers (>500 mW) are hazardous to eye and skin from any exposure: direct, specular, or diffuse (see below). In other words, looking at a spot on the wall from a Class 4 laser can cause eye damage. There is no Class 4R or 4B.

9.2.4.2 Laser beam hazards
Laser beams pose a threat mostly to the eyes. The high power density (or **illuminance**) of the beam, combined with the light-absorptive properties of the eye, make lasers a threat to all portions of the eye. A 1 mW (Class 3R) laser pointer will be focused onto the retina at an irradiance of approximately 100 W/cm^2, about 10 times brighter than the Sun.

Damage to the retina, the cornea, and the vitreous humor of the eye are all possible, depending on the wavelength of the laser and the absorptive properties of the tissue in question. Typically, visible and near-infrared light passes through the cornea and vitreous to reach the retina; burns and bleeding easily produce permanent vision damage by heating the retina faster than local conduction and blood flow can remove the heat—literally cooking the tissue.

IR-B and IR-C light, as well as UV light, is absorbed by the cornea. Higher laser powers can produce permanent burns and clouding through the same mechanism as retinal damage, while lesser powers cause cataract formation later in life.

Non-visible laser wavelengths are considered more hazardous, because the eye has no aversion response to such light. (This is why Class 1 and 2 laser classifications are limited to visible wavelengths only.)

Stronger Class 4 lasers can also threaten the skin and bulk tissue; laser punctures of fingers are not unknown among laser researchers. Class 4 lasers also cause heating of the vitreous humor of the eye, leading to local boiling and "acoustic" damage to eye tissues (e.g., retinal detachment) due to the collapse of vapor bubbles. Victims exposed to these effects frequently report hearing or feeling a "pop" as the beam struck.

Short wavelengths (<600 nm) can also photochemically activate various biomolecules or medications. This is usually undesirable, promoting skin damage and photokeratitis, for example, but it is used in several medical contexts for treatment of specific conditions (such as jaundice in premature infants, for whom gentle, diffuse laser illumination at 475 nm activates bilirubin, a hemoglobin breakdown product, allowing it to be more easily cleared by the liver). The same photo-activation can be detrimental to individuals taking certain medications which act as photosensitizers. Handheld UV lasers are also sometimes employed as photocatalysts for hardening polymer-based dental fillings.

Exposure to laser beams can occur through both **specular** and **diffuse** reflections of the beam, or through **direct viewing**. (See Figure 9.2.) Diffuse reflections occur when a laser beam illuminates a dull surface, such as a wall; such reflections are usually a threat only with Class 4 lasers. Specular reflection occurs off of shiny surfaces: misaligned optical components, watches accidentally inserted into the beam path, etc. Direct viewing usually occurs during laser alignment procedures, when a thoughtless researcher absentmindedly places his or her head in-line with the beam. (The author has done this personally; fortunately, the laser was low-power and the beam passed first through a diffraction grating.) In such cases, the researcher had often removed his or her laser protective eyewear "in order to view the beam better."

9.2.4.3 Non-beam laser hazards

While most researchers pay closest attention to the hazards from the laser beam, lasers actually create a variety of hazards—physical, chemical, and radiation. Taken together, these **non-beam hazards** can pose a greater threat to life and health than the beam. Such hazards include:

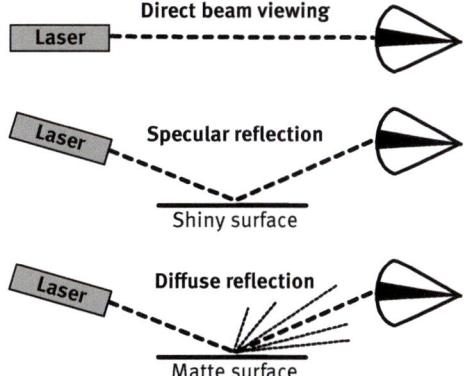

Figure 9.2: Illustration of direct, specular, and diffuse beam viewing.

- **Electricity.** To the author's knowledge, no researcher has ever been killed by a beam exposure; electrocution by high-voltage power supplies has claimed several lives.
- **Fire.** Class 4 lasers are capable of igniting many materials, and a stray beam can easily provide an ignition source for combustible wall coverings, cardboard boxes, etc.
- **Explosion.** Many laser designs require high-voltage discharge lamps as "pumping" agents; such lamps have been known to explode.
- **Chemical hazards.** Dye lasers frequently employ carcinogenic dyes (and for this reason are decreasing in use), while argon-fluorine lasers use dangerous and difficult-to-control elemental fluorine. Other chemical hazards come from the material struck by the laser: ablated materials easily become airborne "LGACs"—Laser-Generated Airborne Contaminants—which can be inhaled to the researcher's detriment. (See Figure 9.3.)
- **Collateral radiation.** Lasers emit wavelengths other than that of the nominal beam wavelength from several sources: pumping lamps or lasers, frequency-doubling optics, and plasmas created by ionizing the air through which the beam passes. Some green laser pointers have been found to emit more IR light than visible [108] due to poor construction and the manner in which the light is produced.

The list above is not exhaustive, [43] contains a more comprehensive listing, and [109] provides a broader introduction to laser safety and laser safety measures.

9.2.4.4 Laser hazard controls

Control of laser beam hazards starts with control of the **Nominal Hazard Zone (NHZ)**, that is, the area in the laboratory within which the beam is bright enough to cause injury. In some installations (especially those which are poorly designed), this may be the entire room, while more clever design of experimental equipment (particularly the optical setup) can reduce the size of this zone considerably.

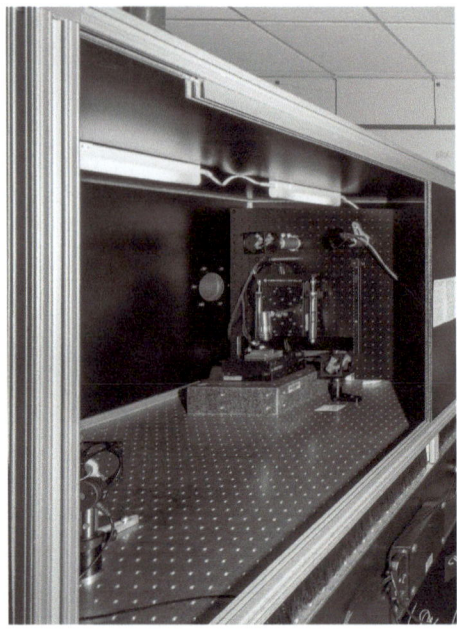

Figure 9.3: Laser micromachining setup. Note the large interlocked beam enclosure that prevents beam exposure.

A combination of engineering controls and administrative controls are used to limit the NHZ and the number of personnel within it. Typical engineering controls include:
- **Laser-opaque beam enclosures or protective housings** to prevent the beam leaving a small controlled area;
- **Key switches** to restrict operation of the laser to authorized personnel;
- **Interlocks** that shut off the laser beam in the event the enclosure (or sometimes the room door) is opened;
- **Special optical laser beam stops**, which absorb the energy of the beam and prevent it from passing beyond a certain point;
- **Warning signs and/or lights** to provide notice that the laser is present and energized;
- **Blackout curtains or walls** (which must be fire-retardant for powerful Class 4 lasers).

Administrative controls for lasers include restricting the area to authorized and trained personnel only and use of detailed Standard Operating Procedures (see box below).

Laser safety standards [43] distinguish between personnel required to work within the NHZ and those who are not permitted to enter. Those working inside the NHZ are typically performing operations such as beam alignment or setup/adjustment of samples. Observation and measurement can often take place outside the NHZ; this considerably simplifies the safety precautions required for such activity.

Personal protective equipment is also necessary for laser exposures. There is an economic incentive to designing a laser installation with as small a NHZ as possible: those working in the NHZ, who may be exposed to hazardous levels of laser energy, **must** be provided with expensive laser protective eyewear, while those outside require no such protection. A typical pair of laser goggles costs 225€ ($250); reducing the number of goggles required can be a major cost savings. Personnel working within the NHZ also require more extensive (and thus more expensive) training.

Because lasers come in a wide variety of powers and wavelengths (sometimes emitting more than one wavelength, even), laser protective eyewear (laser goggles) must be designed for the specific exposure situation. Effective laser goggles must block all laser wavelengths in use, allowing only an eye-safe level of beam energy to penetrate, while still providing enough transparency outside those wavelengths to allow the user to see the room. (Ineffective laser goggles block so much light that the user cannot see the room in order to conduct experiments or see the beam sufficiently to adjust and align it.) Because the analysis methods involved are specialized, it is generally recommended that a laser safety specialist identify the appropriate goggles and specify engineering and administrative controls for any given laser experiment.

Skin protection must also be provided for lasers. For Class 4 lasers, fire-retardant clothing is often used (see section 7.4.2.8 above), while standard lab apparel is appropriate for work with Class 3B lasers. Laser wavelengths in the ultraviolet range may require additional protection—this is one of the rare situations in which sunscreen may be necessary in the lab.

Standard Operating Procedures for equipment
Written procedures are one type of administrative control commonly used in laboratory settings. Such procedures go by several names: **Standard Operating Procedures (SOPs)** are the most common description in industrial settings, while the author has found that academic researchers understand the concept more readily if the SOPs are termed **experimental protocols.**

There are three different types of SOP written for laboratory use. The first are enumerations of general rules of practice for particular (usually hazardous) materials—while common, such "SOPs" are less procedures and more lists of safety rules. The second is specific steps for carrying out a particular experiment or procedure—a true operating procedure—while the third is instructions for operating a particular piece of equipment—also a true procedure. The difference between the last two is a matter of focus and purpose.

SOPs are not simply dry documents written to satisfy regulatory requirements. They have several important uses in the laboratory:
- **They promote consistency** in the conduct of the procedures they describe—which helps to ensure that results are reproducible regardless of which researcher took which data;
- **They serve as excellent training documents** for new members of the laboratory team or as refreshers for researchers who rarely conduct that particular procedure;

- **They provide control over changes** in the way experimental procedures are conducted, as any change must be accompanied by a change to the SOP.

See chapter 18 below for detail on how and when to write SOPs.

SOPs for lasers typically address the specific hazards of the laser in use, PPE requirements, and procedures for required operations—particularly alignment of the laser. One technique commonly used to reduce risk with more powerful lasers is to perform alignment of the experimental optics first with a lower-power Class 2 or 3R laser, only mounting and aligning the Class 3B or 4 laser once the rest of the experiment is fully aligned. This minimizes the amount of time the researcher is exposed to the high-power laser beam, in keeping with the "time-distance-shielding" concept for radiation protection.

9.3 Case studies

9.3.1 Chemical engineering experiment

Electrocatalysis is a burgeoning field that promises advances in energy generation, CO_2 sequestration, and manufacture of high-volume industrial chemicals such as methanol. Improvements in electrochemical catalysts depend critically on understanding of the behavior of various chemical species at the surface of the catalyst. Sum-frequency spectroscopy is a 20-year old laser spectroscopy technique that studies vibrational spectra of chemicals **at an interface**. This specifically interfacial response allows researchers to monitor a functional group or chemical, such as carbon monoxide, as its surroundings change and as a chemical reaction progresses. One particular research group uses sum-frequency spectroscopy to study the pathways a reaction takes on the surface of electrocatalysts, with an eye toward understanding how to better design electrocatalysts.

The research group uses a solid-state titanium-sapphire "passively mode-locked" pulsed laser with a pulse duration of about 1 ps at 40 Hz and a power of about 40 mJ/pulse. (This is easily a Class 4 laser.) The short pulse duration can pose a threat to normal laser protective goggles, photobleaching them and bypassing their protection, so caution is necessary in the design of the laser setup and in the selection of appropriately-rated laser protective eyewear. The fundamental frequency of the laser is 1064 nm, in the near-infrared range, but sum-frequency spectroscopy requires several other beams, all synchronized in space and time so that they strike the sample simultaneously.

The additional beams are obtained by passing the fundamental through several non-linear optical crystals, producing a set of beams at 1064 nm (8 mJ), 355 nm (near-UV, 7 mJ), and 532 nm (green, 100 µJ). The low-power visible beam is a probe beam, while the others drive nonlinear optical parametric amplifiers and optical parametric generators which produce a **tunable** beam in the mid-IR region of 2300–1100 nm. (See Figure 9.4.)

Figure 9.4: Partial beamline of sum-frequency laser spectrometer.

These beams illuminate an electrochemical cell. As mentioned above, the beams must overlap both temporally and spatially. Second-order nonlinear processes at the surface generate "sum-frequency" photons, which (not surprisingly) are the sum of the mid-IR exciter beam and the 532 nm probe beam. These photons are collected and a spectrum obtained. Figure 9.5 shows a researcher working in the sample area to collect a spectrum.

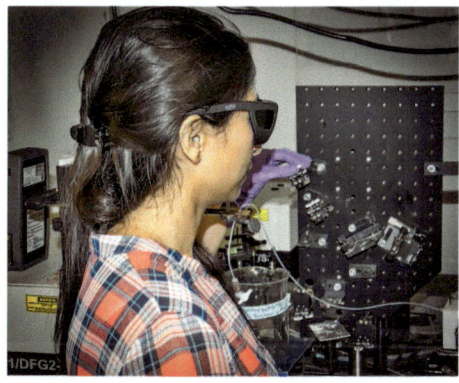

Figure 9.5: Dr Shalaka Dewan operates a sum-frequency laser spectrometer.

This experiment's hazards center around the various laser beams. Several features complicate the control of hazards. First among these is the variety of wavelengths in use: laser beams ranging from 11 000 nm to 355 nm are used at various points, and eye protection (as well as beam stops and other components) must protect against all of these wavelengths. In addition, the ultrashort picosecond pulses can photochemically bleach eyewear, destroying their protection, at least for a time.

The keys to protection from the laser beams in this installation are to maintain the laser beam in a single plane as much as possible and to provide appropriate laser protective eyewear. If the beam is kept in a single plane, it is far less likely that a stray hand or tool will find its way into the beam and be damaged or reflect the beam in a random direction. In this case, it is not possible to keep the beam entirely horizontal, because the exciter and probe beams must be directed onto the sample along a path with a vertical component (the sample cannot be turned on its side because it is liquid-filled and unsealed). To reduce the probability of random reflections, the sample area is surrounded on three sides by beam-blocking materials.

Appropriate eyewear is a challenge in this experiment. The 532 nm beam is visible and has only 100 µJ of energy, and so poses much less of a threat; the mid-IR beam is several mJ in power and cannot be seen. The third-harmonic used in generating the mid-IR beam, 7 mJ at 355 nm, is in the near-UV region. This means that safety eyewear must block (as completely as possible) the IR and UV regions while allowing visible light transmittance to be as high as possible (so that the researcher can see to work). Specially-rated eyewear constructed to resist photobleaching from picosecond pulses may also be appropriate.

9.3.2 Medical experiment

Any surgical intervention is an inherently difficult and dangerous procedure, for both patient and surgeon. The patient bears the majority of the risks of the intervention, with the concomitant physical and emotional difficulties, while the surgeon must negotiate his or her way through often-cramped anatomical constructs, deal with complications, and may be exposed to a variety of pathogens or other dangerous materials during the course of an operation. In orthopedic surgery, the procedure is often guided by X-ray imaging using a C-arm X-ray machine (see Figure 9.6); placing a single Kirschner wire (an orthopedic "pin") can require many tens of images, resulting in a high radiation dose both for the patient (whose tissue is being imaged) and the surgeon (whose hands are in the X-ray exposure zone).

One research group is experimenting with use of augmented-reality (or "mixed reality") imaging systems to reduce the procedure time, improve accuracy, and reduce radiation dose to surgeon and patient. This is accomplished through superposition of a 3-D visual image of the patient with a computed radiological image (generated by a technique called **cone-beam computed tomography**, CBCT) [110].

There are several radiation hazards in this experimentation. Direct radiation exposure to the patient and the surgeon is the most obvious. Nevertheless, a significant amount of radiation scatters from the patient's body and the support on which he or she rests (see Figure 9.7), adding exposure to the surgeon in particular. The laboratory in which this work is conducted is a demonstration operating room, and others could be working in the room during an experiment. Beyond the surgeon, several other experimental personnel must be present to operate the augmented-reality system and observe the surgeon's performance; all of these personnel are potentially exposed to X-ray scatter from the C-arm. Finally, the room is a showplace laboratory, with one wall (the one nearest the C-arm) being made of windows opening on a public throughway; public radiation exposure is a concern as well.

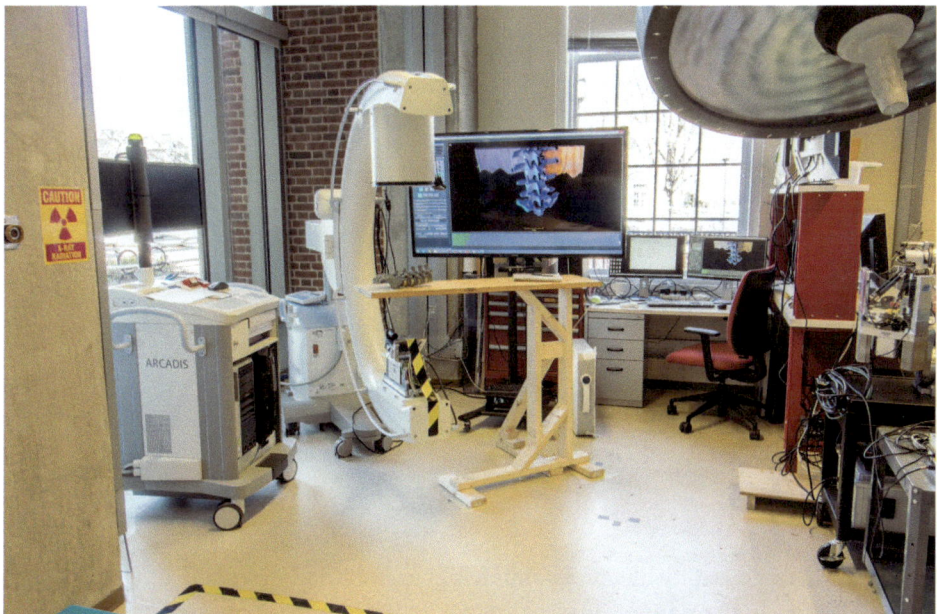

Figure 9.6: C-arm X-ray machine.

The C-arm X-ray machine is already engineered to reduce scatter and extraneous exposure as much as possible, and provision of additional engineering controls (such as leaded glass windows to protect passers-by) would be difficult. The lab therefore relies on a mixture of strong administrative controls and personal protective equipment to ensure radiation safety. No personnel not associated with the experiment in progress are allowed in the room, and work is done at night, when there is little or no traffic on the public throughway. Health physicists monitor exposure to laboratory personnel, and radiation dosimeters are mounted on the windows to detect any public exposure (none has ever been detected). Experimenters and surgeons wear heavily-leaded aprons and other gear to protect radiologically-sensitive organs, and experimenters stand at a substantial distance from the X-ray machine that reduces their radiation dose to nearly zero. The research is at a stage where work may be done using polyurethane "phantoms," so no patients are exposed to risk during the experiments.

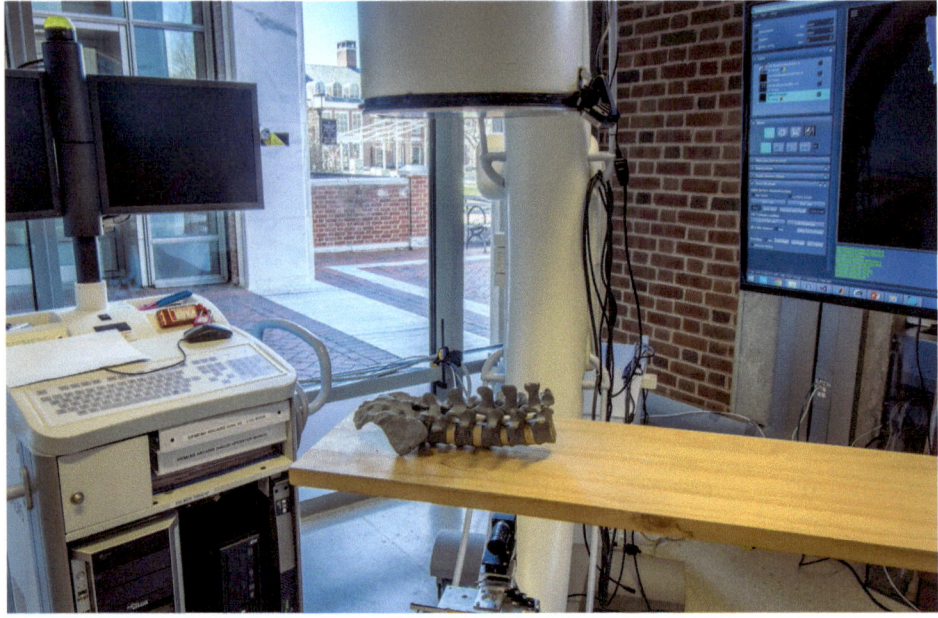

Figure 9.7: Anatomical model on stage of C-arm X-ray machine.

Oddly, CBCT radiography involves a much higher radiation dose than a typical C-arm image. Assuming the patient does not move, though, images need only be taken at the start and end of the procedure instead of constantly throughout. This results in a much lower radiation dose for the surgeon (and presumably for the patient, although this has not yet been measured.)

9.3.3 Exercises: radiation hazards

1. Answer the following questions:
    A. True or False? Laser protective goggles are required for work with Class 3R lasers.
    B. True or False? Ultraviolet light exposure can cause cataracts.
    C. True or False? Protection from low-energy beta-emitting radioisotopes such as tritium can be achieved by a few centimeters of some plastics.

D. True or False? Researcher exposure to ionizing radiation should be reduced As Low as Reasonably Convenient.
   E. True or False? Laser beams are the greatest threat to life from lasers.
2. Examine the following brief laboratory radiation-use situations and give at least one practicable control to reduce risk from the hazard.
   A. A researcher is measuring turbulence in a fluid sample using laser interferometry employing a Class 4 laser. The experiment itself is confined to a single optical table.
   B. A civil engineer is measuring the density of a concrete sample using a **nuclear densometer**, which operates by inserting a γ-radiation source (^{137}Cs) into a hole drilled in the sample and measuring gamma counts at the surface. The measurement takes about 60 seconds.
   C. A biologist is conducting electrophoresis measurements of nucleic acid samples—so-called "DNA fingerprinting"—employing ethidium bromide as a DNA stain. Ethidum bromide absorbs UV light at 285 nm, emitting a visible orange glow. The results are visualized using a **UV transilluminator**, an instrument for shining a strong UV light through a gel from below.

Part III: Hazard analysis techniques

10 The Checklist technique

The **Checklist technique** is the simplest form of hazard analysis available, and it is exactly as it sounds: the user sequentially follows a checklist of hazards and/or scenarios.

The form of the checklist varies. Some checklists are simple listings of hazards common to a situation, technology, or operation; in these cases, the purpose of the checklist is to help the analyst determine that all hazards frequently found in such places are identified, leaving the exact method of control up to the user. Chapter 16 below presents a checklist of this kind—it is intended to help identify general classes of hazard that may be present in a newly-proposed experiment. (The author uses a variant of this checklist for reviewing experiments and for screening new research programs for early identification of hazards.)

Other checklists for highly-defined, common situations such as delivering gas from a compressed gas cylinder (see section 17.1 below). Such lists are worded in the form of questions about the hazard controls required, and the user is meant to ensure that each item is either in place or not relevant to the situation.

Checklists are used outside the lab in a variety of situations, such as aviation and medical clinical practice [41, 111]. In these applications, checklists often ensure that all steps of a procedure are performed in the correct order.

10.1 Strengths, weaknesses, and suitability

The main strength of the checklist approach is that it ensures that one covers the most common hazards. This assumes, though, that the checklist is comprehensive.

The most prominent weakness of checklists is that they are often assumed by the user actually to **be** comprehensive—leading the user to a false sense of security. A checklist can divert attention from obvious hazards that are **not** on the list.

For this reason, checklists alone should only be used for highly controlled situations that occur regularly and whose hazards are well-understood. In general, they are more useful in combination with other techniques such as What-If? (see chapter 12 below) or JHA (see chapter 11 below), providing a baseline from which to begin and assuring a minimum level of coverage.

10.2 Sources of checklists

Checklists may be obtained from a variety of sources. Scientific associations such as the American Chemical Society or the American Society of Physics Teachers publish checklists for various types of operations (see, e.g., [112]), and trade associations such as the (US) Compressed Gas Association publish standards that provide excellent raw material for new checklists.

A set of sample checklists for common laboratory operations is presented in chapter 17 below for the use of the reader. Do not assume that they contain address all significant risks in your particular use; use them as starting points for a more thoughtful analysis of your activities.

10.3 Example checklist: Quick laboratory inspection

The checklist below may be used for quick inspection of physical conditions in a typical laboratory—it is meant as a short example and is by no means comprehensive. The reader may wish to use it as the starting point for a laboratory inspection checklist suitable for his or her lab.

As a matter of good practice, the inspection checklist is worded in the form of questions about the lab; in a "perfect lab," all questions would be answered in the affirmative. See Table 10.1 for an example.

Table 10.1: Sample checklist review for lab inspection

Question	Yes/No	Action to be taken
Equipment		
Is the eyewash in working order?		
Is the safety shower in working order?		
Is access to the eyewash and safety shower unimpeded?		
Does the laboratory have an appropriate fire extinguisher?		
Is the fire extinguisher not outdated?		
Are the fire doors unblocked and normally closed?		
Storage		
Are all flammable solids and liquids stored in approved flammables cabinets when not in-use?		
Are all corrosive materials stored in appropriate corrosives storage cabinets when not in-use?		
Are poisons and other high-hazard chemicals stored in locked cabinets or otherwise secured?		
Is the floor clear of stored items?		
Are the benches clear of stored items?		

10.3 Example checklist: Quick laboratory inspection

Table 10.1: (continued)

Question	Yes/No	Action to be taken
Laboratory practice		
Is food and drink not present in the lab?		
Is a place for food and drink provided outside the lab?		
Are trash cans, biohazardous waste containers, and sharps containers not filled to overflowing?		
Is appropriate personal protective equipment (e.g., lab coat, safety goggles) provided?		
Is personal protective equipment appropriately stored when not in-use?		
Compressed gases		
Are all cylinders properly secured?		
Are cylinder caps in place when cylinders are not in-use?		
Are appropriate regulators (e.g., compatible with the gas, of the appropriate pressure delivery range, etc.) available or mounted?		
Are highly toxic or otherwise dangerous compounds (e.g., silane) stored and used in ventilated gas cabinets?		

Table 10.2 is an example of a completed Checklist review using the checklist above.

Table 10.2: Completed checklist review from Table 10.1

Question	Yes/No	Action to be taken
Equipment		
Is the eyewash in working order?	Y	
Is the safety shower in working order?	?	Request Facilities Management to test safety shower.
Is access to the eyewash and safety shower unimpeded?	N	Remove cylinders from underneath safety shower.
Does the laboratory have an appropriate fire extinguisher?	Y	
Is the fire extinguisher not outdated?	Y	
Are the fire doors unblocked and normally closed?	Y	

Table 10.2: (continued)

Question	Yes/No	Action to be taken
Storage		
Are all flammable solids and liquids stored in approved flammables cabinets when not in-use?	N	Place all unused flammables in storage.
Are all corrosive materials stored in appropriate corrosives storage cabinets when not in-use?	Y	
Are poisons and other high-hazard chemicals stored in locked cabinets or otherwise secured?	N	Purchase and install cabinet locks for poisons cabinet.
Is the floor clear of stored items?	N	Move cases of gloves on floor to the storage room.
Are the benches clear of stored items?	Y	
Laboratory practice		
Is food and drink not present in the lab?	N	Evidence of food and drink present (coffee cup).
Is a place for food and drink provided outside the lab?	N	Install a shelf in the hallway outside the lab for coffee cups.
Are trash cans, biohazardous waste containers, and sharps containers not filled to overflowing?	Y	
Is appropriate personal protective equipment (e.g., lab coat, safety goggles) provided?	N	Provide safety goggles instead of safety glasses for this laboratory.
Is personal protective equipment appropriately stored when not in-use?	N	Do not hang lab coats from the top of gas cylinders.
Compressed gases		
Are all cylinders properly secured?	Y	
Are cylinder caps in place when cylinders are not in-use?	Y	
Are appropriate regulators (e.g., compatible with the gas, of the appropriate pressure delivery range, etc.) available or mounted?	N	Regulator on nitrogen purge gas cylinder can deliver 500 kPa while instrument and delivery tubing can withstand only 200 kPa. Replace regulator with appropriate delivery range.
Are highly toxic or otherwise dangerous compounds (e.g., silane) stored and used in ventilated gas cabinets?	N/A	No highly-toxic/pyrophoric/otherwise dangerous compounds stored in this lab.

10.4 Evaluating recommendations from hazard analyses

The purpose of hazard analysis is to identify hazard scenarios, estimate the risk they present, and optionally make recommendations for reducing those risks. It is **not** to make detailed consideration of cost, effectiveness, or practicality—it is not primarily for solving problems identified in the analysis. The point is to identify problems that pose risks, **not to conclusively solve them**. A hazard analysis should, at a minimum, identify those risks in an experiment or facility that require or deserve attention to reduce them.

This is not to say that many hazard analysis method do not make recommendations for exactly how to reduce risk; it is common to make at least an initial attempt at solving the risk reduction problems posed. It is essential, nevertheless, to concentrate on **identifying hazard scenarios and gauging the risks first** and making recommendations for risk reduction second.

When a hazard analysis makes specific "recommendations," it should be for one of two reasons: either the solution to the issue is generally agreed (e.g., the risk of a gas cylinder falling over and becoming damaged is almost always controlled by restraining the cylinder) or legally required (e.g., placing containers of flammable liquids larger than 4 L into an appropriate storage cabinet). Checklists are often constructed to guide the user to these generally-accepted solutions through wording such as "Is the pressure safety relief valve tested monthly?"

All other "recommendations" should be considered preliminary—there are usually other ways to reduce the risk, and they may or may not be the appropriate long-term solution. The author suggests wording recommendations in the form of suggestions to reduce the risk of a particular scenario ("consider reducing the risk of flying shrapnel from shattered samples") instead of prescribing particular solutions ("place a 2 cm polycarbonate guard around the apparatus") unless the solution is required or clearly obvious.

10.5 Exercises: laboratory inspection

1. Expand the checklist in Table 10.1 to include questions specifically appropriate to your discipline.
2. Inspect an operating laboratory using the checklist prepared in Exercise 1 above.

11 The Job Hazard Analysis technique (JHA)

Job Hazard Analysis (JHA), also called **Job Safety Analysis (JSA)**, decomposes a task into a simpler series of subtasks, evaluating hazards and appropriate controls for each subtask.

11.1 Strengths, weaknesses, and suitability

JHA is well-suited to analysis of simple, especially repeating, tasks in a variety of fields ranging from construction to chemistry. Its greatest strength is that is it simple enough to be workable by less-experienced personnel (e.g., a construction-site foreman or an early-stage graduate researcher). In particular, the JHA technique scales well to both the scope of the analysis and the sophistication of the analyst: a task may be broken down into 25 detailed steps or 5 broader ones, and a Job Hazard Analysis may be performed on either one. This allows the depth of the analysis to be varied depending upon the expertise of the staff that will conduct the task and on the resources available for the analysis.

Nevertheless, while task decomposition is relatively systematic, the actual hazard identification and control is somewhat **ad hoc** and prone to omission and error. A standard Job Hazard Analysis mixes the concepts of **hazard** and **hazard scenario** (see section 4.1.1 above) and relegates risk estimation to an implicit judgment without any hint of objectivity.

JHA may be used for simpler laboratory protocols (e.g., DNA staining and visualization), standardized operations for equipment (e.g., sharpening a tool, setting up a model in a wind tunnel), and it is also suitable for task or equipment design, in the form of a related technique (not covered here) called Preliminary Hazard Analysis.

11.2 Technique

To conduct a Job Hazard Analysis, the analyst first decomposes the task under study into manageable steps of a scope appropriate to the analysis. (Refer to the next section for an example.) The complexity of the steps (the **granularity** of the analysis) must be chosen to be appropriate to the task, the information available about its hazards, the skill of those who will carry out the task, and how much time and effort the analysis has available for the analysis. A JHA can be carried out in the office of a safety professional, with great attention paid to the possible hazards of each task, or in the field by a construction supervisor or plumber, as a quick, simple check that the most important and obvious hazards of the tasks have been addressed.

After task decomposition, the "hazards" of each step are listed. In a JHA, the word "hazards" includes the hazardous properties of the task—flammable chemicals, for example, or sharp edges on a freshly-cut steel sheet—but it also includes the means by which exposure to the hazard (and consequent injury or death) might occur. Flammable chemicals might be ignited by a nearby water heater, for example, or a sharp-edged steel sheet might cause cuts to the hands of the worker handling it—or, at worst, cause amputation of a digit. The risk in a JHA is that the analysis may not be fully comprehensive at this stage, omitting fairly obvious hazards because of the analyst's lack of expertise and experience with the task (or because of the inability to spend an appropriate amount of time and thought on the analysis).

Once the hazards are known, all known hazard controls applicable to the hazards of that step should be listed. Heavy leather work gloves that protect against injury from sharp metal edges are an example, as is a modern sealed-combustion water heater that is unlikely to ignite flammable vapors.

At this point, the analyst must make a risk judgement: are the controls provided adequate to protect against the hazards identified? Unfortunately, "Risk" is not normally a column on a Job Hazard Analysis worksheet. While semi-quantitative methods for making this judgement are available, most of the time risk estimation, and consequently the decision about whether the level of risk remaining is acceptable is done in an **ad hoc** manner—representing the analyst's gut feeling" instead of some defensible, repeatable method. At the very least, a JHA should include some record of an estimated qualitative level of risk (perhaps "High" for obviously-unacceptable risks, "Medium" for borderline cases, and "Low" for acceptable risks), and preferably a short note as to **why** the risk was evaluated in that manner.

If the risk of exposure to the hazard(s) for the step being studied is judged unacceptable, additional controls should be prescribed. Often, the analyst will specify additional controls applicable to those hazard(s) that are not adequately controlled at present. Such rigid specifications are appropriate only when the analyst is the decision maker responsible for implementing the controls.

As mentioned in section 10.4 above, it can be useful (and is almost necessary in more complicated methods of hazard analysis) to separate the issue of judging the existing level of risk to be acceptable or unacceptable and that of prescribing additional controls as required solutions to bring the excessive risks to a tolerable level. This allows the hazard analysis (which might be conducted by a foreman or graduate student) to concentrate on identifying problems, leaving the finding of solutions for higher-level personnel (e.g., site manager, principal investigator). In this case, the "additional controls" might not be explicitly specified—instead a notation would be made that attention should be given to reducing the risk of the unacceptably-risky hazards identified.

11.3 Example JHA

Consider a mechanical engineer manufacturing prototype parts in a machine shop. The task might be broken down into several steps: gathering materials, using a laser cutter to cut the basic shape, using a drill press to insert holes in appropriate places, employing a brake (metal-bending machine tool) to create precise bends, washing the part in a solvent washer, and dip-coating the part in a solvent-based paint. A simple JHA worksheet for this task might look like Table 11.1:

Table 11.1: Example Job Hazard Analysis worksheet

#	Step	Hazard(s)	Control(s)
1	Gather materials	Thin steel stock is sharp and may cause cuts.	Wear heavy leather gloves
2	Cut with laser cutter	Freshly-cut steel is even more likely to have sharp edges. Laser cutter generates exhaust laden with ablated metal.	Wear heavy leather gloves, Turn cutter exhaust ON before use
3	Drill with drill press	Drill turnings may get into eyes. Drilling fluid may splash on skin or in eyes	Wear impact-resistant splash goggles Wear shop jumpsuit to protect against splashes.
4	Bend with brake	Pinch point on brake	Persons using brake MUST be trained and qualified by shop supervisor
5	Solvent-wash part	Solvent (Methyl ethyl ketone) is toxic	Do not put head into chamber of solvent washer. Wear solvent-resistant gloves
6	Dip-coat part	Solvent-borne coating gives off carcinogenic vapors (methylene chloride) as it dries Coating may splash	Dip-coating and drying MUST be conducted only in ventilated paint room. Wear solvent-resistant gloves and splash goggles. Wear plastic-coated disposable painting suit to protect against splashes.

11.4 Exercises: Job Hazard Analysis

1. Suppose the risk of exposure to carcinogenic solvent during the dip coating step in Table 11.1 is still considered too high. What additional measures could be taken to control the risk?

2. Consider an operation to propagate cell cultures of human-derived cancer cells considered biohazardous. The operation is conducted weekly in order to maintain a viable culture, and it can be decomposed into the following steps:
   A. Remove culture flask from (37°C) incubator filled with CO_2.
   B. Verify culture viability by inspection under microscope to ensure cell count exceeds 1,000 cells/mL.
   C. Prepare new culture flask containing culture medium (which contains toxic components but is available commercially).
   D. Transfer 5 mL of old cell culture to new flask under sterile conditions.
   E. Seal and place new culture flask in incubator.
   F. Destroy old culture by mixing with 5% bleach in water, allowing 30 min sanitization time, and disposing in sink.
   G. Dispose of all contaminated materials by autoclaving, then placing in biohazard box marked "pathogenic".

   Conduct a JHA on this operation.
3. Imagine you have experienced a blown-out tire on the highway. Decompose the operation of pulling over, changing the tire, and returning to the highway into a list of tasks at an appropriate level of complexity for a typical driver, then conduct a Job Hazard Analysis on the task.

12 The What-If? technique

The **What-If? Technique** permits a more in-depth analysis of the safety status of a particular procedure, apparatus, or experiment. What-If? is more broadly applicable than the Checklist or JHA methodologies, and it is, when performed correctly, more comprehensive than either.

The What-If? Technique begins with a question or condition that may be inherently hazardous or that is thought potentially to lead to hazardous situations. The name of the method comes from the fact that such questions often begin "What if…"? (As in "What if the microtome blade breaks during cutting?") This phrasing is not essential, however; variations and evolutions on the What-If technique (such as the Hazard and Operability Study, which is commonly used in the chemical industry) use other methods to develop potential hazardous conditions for study.

12.1 Strengths and weaknesses

The What-If? technique has the advantage of offering a moderately-complete analysis of each question posed. Techniques such as Checklist can easily omit hazards not envisioned when the checklist was made, while Job Hazard Analysis is reliant upon intuition to identify hazards and appropriate controls. The technique imposes more structure on the analysis, while still allowing room for creativity and expertise.

Depending on the variant used, the What-If? technique does suffer from the disadvantage that the actual "What-If?" questions are produced by brainstorming or other similar creative processes. This approach is vulnerable to the analysts' lack of knowledge in particular areas (e.g., an analysis team without a mechanical engineer is unlikely to identify structural issues in pressure design of equipment) and poor group dynamics (e.g., group members unwilling to speak up and be assertive—this is a particular problem in workplaces such as academic research groups where a strong hierarchy and "social pecking order" exists).

12.2 Suitability

In the laboratory, the What-If? technique is appropriate for equipment and/or operations of moderate complexity, more complicated than what could typically be expected to be studied using a simple JHA. A new major piece of laboratory equipment (such as a laser ablation apparatus or mechanical tester), or the operating procedures for a hydrogenation reactor might be appropriate topics for a What-If? technique analysis.

It is also suitable for situations that are potentially more risky, as What-If? technique studies the situation to a greater level of detail and much more systematically.

Sometimes during a JHA or Checklist study, steps or situations are developed which obviously carry high levels of risk or are unclear about the risk level posed. A portion of an operating procedure that carries potentially-lethal consequences if performed out-of-order (e.g., drying ones hands **before** unplugging electrical components) could be studied in more detail using the What-If? technique.

The What-If technique need not be used for the entire process, equipment, or procedure being examined. It is perfectly acceptable to use a checklist for the simpler, more standardized portions of an experiment (setup, chemical disposal, etc.), switching to the What-If? technique for the novel portions of the work.

12.3 What-If? Technique

The term What-If? technique (or What-If? study) covers several variations on the same theme—studying the causes and consequences of various hypothetical situations in a system. The description below details the author's typical method of practice, which tends toward the rigorous. It is very close to that used in the Hazard and Operability (HAZOP) study, although HAZOP employs specially structured methods to generate questions (or in its case, conditions for study).

Because the What-If? Technique generates a moderately large amount of analysis data, special-purpose database applications are available to assist in tracking and documenting the study. These applications are most appropriate for complex industrial systems or for situations in which complete documentation must be kept for regulatory reasons. For laboratory purposes, a simple text worksheet (see Table 12.1) will suffice.

12.3.1 Scoping

The first task to be performed in a What-If? Technique hazard study is to determine the purpose and scope of the analysis: what system or parts of a system are to be studied, and what conditions are to be assumed. For example, one might decide to study a new gas chromatograph/mass spectrometer in its entirety, including the supporting utilities and other facilities, or one might restrict study to the effects of utilities failure on the system.

Usually, the system is divided into smaller sections ("nodes") for convenience and organization. The convention typically is that while analyzing a node, only questions or conditions originating within the node will be studied, but that the implications of the question will be followed into other nodes as far as necessary. It is recommended to define a node for "external causes" outside the formal scope of the analysis to provide a place for such questions as "What if the basement laboratory floods?" as such conditions do not originate specifically with any node of the system under study.

Table 12.1: What-If? Worksheet

What-If? Worksheet Page X of Y

Date
Node/System section
Study team

What-If?	Causes	Consequences	Controls	Initial risk	Recommendations/Actions (Assigned to)	Final risk

12.3.2 Team assembly

A What-If? analysis typically employs more than one person, although it can be conducted alone. Involvement of multiple personnel allows a broader range of expertise and experience to be brought to bear. As mentioned above, if mechanical integrity issues are potential problems, inclusion of a mechanical engineer in the study team would be prudent. The author has found that scientific and engineering investigators tend to focus on hazards in their particular areas of expertise: that is, chemists will readily identify chemical hazards, mechanical engineers will identify physical/mechanical hazards most easily, etc.

Assemble a varied team of experts, including as necessary the principal investigator, other more experienced research personnel (senior graduate students, post-doctoral fellows, staff engineers, etc.), perhaps a facilities representative or building engineer for facilities-dependent systems. **Including a technical professional (graduate student, engineering technician, etc.)** who has conducted similar work with similar equipment before is **strongly** recommended, as is inclusion of a manufacturer's representative for large, expensive installations (e.g., magnetic resonance imager).

12.3.3 What-If?

The first step in a What-If? study is to pose relevant questions for study. Often these are hypothetical situations, worded to begin, "What if...?", such as "What if the cooling water fails halfway through the experiment?" or "What if the distillation column is accidentally charged with methanol instead of ethanol?"

Asking appropriate What-If? questions is a highly creative activity, and team members (or individual analysts) should use any creativity-promoting method they prefer to assist. For example, the team may choose to begin with a brainstorming session to identify as many What-If? questions as possible before beginning the analysis. "Guided brainstorming" in which question generation is assisted through the use of guide words or phrases such as "doesn't happen" or "too much" can be used; one variant on this is termed the Structured What-If Technique (SWIFT) [113].

The What-If? study requires the team or analyst to switch between creative modes of thought—used for identifying questions for study and recommendations for addressing identified risks—and analytical modes. The author has found that more-educated study team members have more difficulty with this "context changing," and that special effort is required to overcome it. The effect may be partially due to aging—as we become older and more experienced, we tend to develop mental heuristics to quickly decide what is possible and what is impossible; this detracts from creative thought. It may also be a function of the methods used for scientific and engineering education: throughout our education, but particularly at the undergraduate level, pedagogy concentrates on methods to solve problems, rather than methods to identify problems and decide what to do about them. In any event, persons conducting hazard analyses should be aware of the bias and take positive steps to counter it.

The steps described in sections 12.3.4–12.3.9 should be followed for each What-If? question according to the algorithm sketched in Figure 12.1.

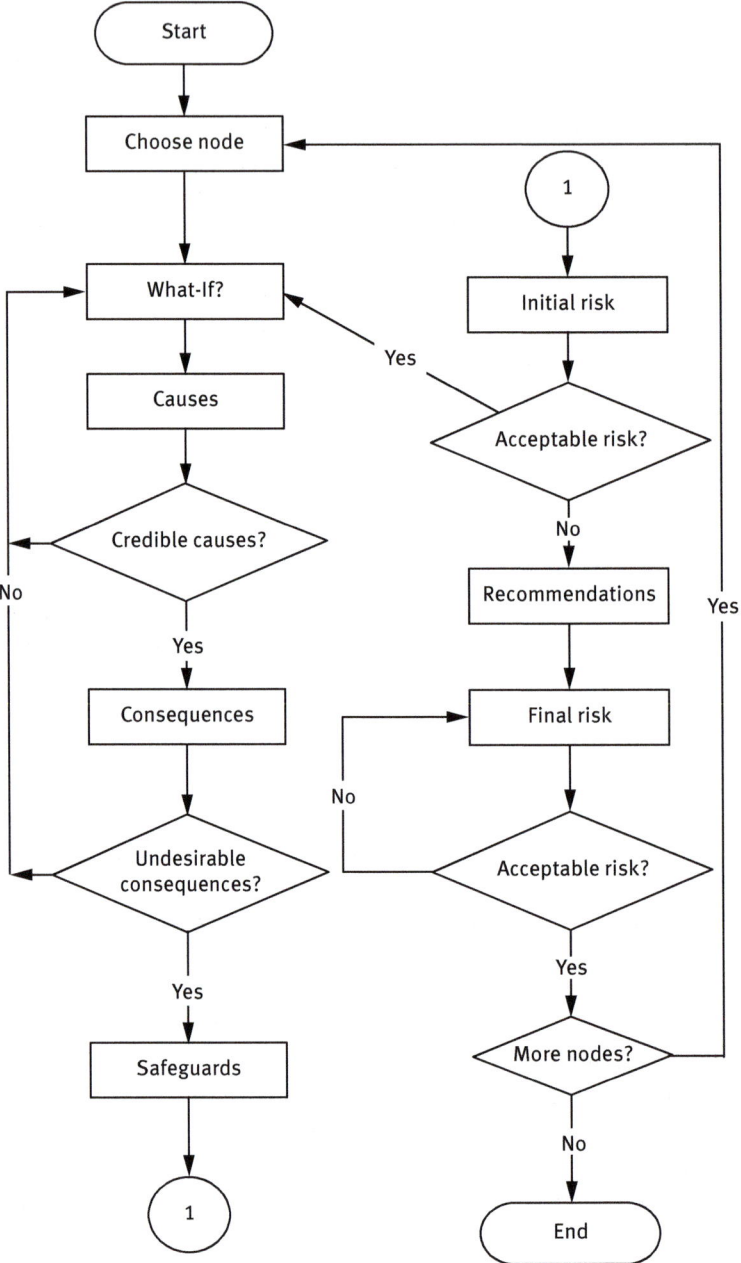

Figure 12.1: What-If? technique algorithm.

12.3.4 Causes

To begin, the situation described by the What-If? question is analyzed for causes: what events or conditions can cause the situation described? As mentioned above, **typically only causes within the current node should be considered**. Causes arising from other nodes will be developed naturally as consequences of conditions occurring in those nodes, so concentrating the analysis on causes within the node avoids "double-counting" and extra, unnecessary work.

If the team were studying a distillation column, for example, possible causes for "What if the column is charged with methanol instead of ethanol?" might be "Purchasing department erroneously orders **methyl** alcohol instead of **ethyl** alcohol," or "Supplier ships incorrect feedstock chemical," or "Methanol left in still after previous run with that chemical."

After the possible causes of the What-If? are enumerated, the team or analyst must make a judgment: **are any of the listed causes plausible**? That is, is it readily imaginable that one more of them might actually occur in the system? If the team can find no plausible causes, the analysis of that What-If? question should be abandoned and study of the next question begun.

12.3.5 Consequences

If there are plausible causes for the situation, the possible consequences should be explored—**What happens if the situation occurs or one of the causes comes to pass?** In contrast to analyzing causes, the consequences of the What-If? question should be carried out of the node as far as practical. This allows identification of **consequences** that would be considered **causes** of events in other nodes.

The most important factor in analyzing consequences of a situation, whether in an industrial or a research setting, is to continue asking oneself, **"What happens next?"** Do not stop with a single consequence—ask what the **follow-on consequences** are. As an example: if a person fills a revolver with cartridges, aims it at another person, cocks the hammer, and pulls the trigger, what happens? If only immediate consequences are considered, the answer is "the hammer falls," and if follow-on consequences are ignored, one gets the erroneous impression that this situation is nonhazardous. In actuality, the fall of the hammer causes the hammer to strike the end of the cartridge; the primer ignites from the force of the impact; the powder charge combusts, generating gas; the pressure of the gas causes the bullet to be forced at high speed down the revolver barrel, and if the shooter's aim is good enough, the bullet strikes the victim, causing injury or death. This series of consequences is clearly **not innocuous**, but the What-If? analysis would fail to identify it if the analyst failed to continue asking "What happens next?"

Note that a single consequence may lead to multiple follow-on consequences instead of just a linear sequence, and each branch should be explored. Water overflowing a tank may initially flow onto the laboratory floor, but from there it might flow into a floor drain, it might destroy boxes of materials stored on the floor, or it might flow into an electrical floor outlet and create a serious shock hazard for a person stepping into the puddle. **Do not stop listing consequences until all branches are fully explored, to the point that the situation is known to be stable and safe.**

If no undesirable consequences are identified, the study team or analyst should move to the next What-If? question; if "bad things happen," continue the analysis.

12.3.6 Controls

Once the consequences of the What-If? question are known, identify what hazard controls **currently exist** or are already in-place in the system. Engineering controls, administrative controls, and personal protective equipment should all be considered. A useful method for identifying controls is to examine each cause and consequence, asking, "What controls are in place to prevent, mitigate, or detect this situation?" In the case of the distillation column, there might be pre-existing (administrative) quality control checks done on the materials to be distilled that would distinguish methanol from ethanol.

12.3.7 Current risk

Following the identification of causes, consequences, and controls, the analyst or study team must make a judgment: **are the risks created by this "What-If?" question adequately controlled?** That is, is the aggregate risk within tolerable limits or are there causes or consequences that do not have sufficient controls? This judgment may be made by any method from gut instinct to formal risk estimation methods such as those described in section 3.2 above.

It is important to consider all causes and consequences together—consider each cause-consequence pair to be a "pathway" to a potential incident. If insufficient controls exist to block that pathway, then the residual risk may be unacceptable and should be evaluated. It is very common to have controls that apply to only a subset of causes or consequences, or controls whose reliability is questionable. This particularly occurs with hazard controls that detect hazardous conditions or accidents in progress, both because detectors are often complex, sensitive instruments requiring frequent, skilled maintenance, or because in order to be effective hazard controls, detectors must be paired with planning and training regarding what actions should be taken when the detector sounds an alarm.

One common method the author uses in lab is a "semiquantitative instinct" risk rating. The risk is evaluated qualitatively and rated **High, Medium, or Low.** A High rating would indicate that the risk is clearly intolerable and **must be addressed** before conducting the experiment; if methanol were a common chemical in the lab and lab workers (e.g., undergraduate students) were unsophisticated enough that mistaken use of methanol in place of ethanol might plausibly be expected, the risk might be High. Medium risk denotes a situation that **should be addressed** provided time, money, and other resources are available, but could be accepted if necessary. Relying solely on purchasing specifications and a manufacturer's certificate of analysis to determine the chemical identity of the alcohol used might be a Medium risk, if the consequences were severe enough. A Low risk is immediately recognizable as tolerable or insignificant: If charging methanol to the distillation column causes only off-spec product of low value and would be immediately recognized by the operator of the still once the distillation began (perhaps because boiling temperatures would be much lower), then the risk might be rated Low.

If the risk of the What-If? question being studied is tolerable or Low, no further analysis is necessary and the study should move to the next question.

12.3.8 Recommendations

If a risk is judged to require further reduction beyond that afforded by the currently-existing controls or safeguards, recommendations should be made for effecting that risk reduction.

There are two schools of thought on the focus of such recommendations. One point of view is that the purpose of any hazard analysis is to identify hazards, not to become sidetracked with detailed analysis of what controls should be added; the other is that specific ideas for risk reduction that occur to the team should be at least mentioned as recommendations. In the first case, the only "recommendation" made would be to take steps to reduce the risk of a particular scenario (e.g., the methanol-ethanol substitution), although the recommendation might be expanded to specifically call out causes or consequences that are inadequately controlled. In the second case, concrete advice is provided for controlling the risk further (e.g., verifying the manufacturer's certificate of analysis by gas chromatography/mass spectrometry—GC/MS—before using the chemical).

While it may be tempting to make very detailed recommendations to control the risk identified, there can be serious drawbacks. If a system is sufficiently risky as to justify the effort of a What-If? study, then the findings and recommendations should probably be reviewed by more senior personnel—in industry this would be a manager—the engineer in responsible charge or the scientist in charge of the division, while in academic labs, the responsible person is generally the principal investigator. Wording recommendations in too narrow a manner constrains the ability of the reviewer to find alternative solutions to the problems identified by the study (see section 10.4 above).

The author's position is that specific recommendations (those which "solve the problem" as opposed to those that recommend solving the problem) should be made only if the solution to the problem is either simple, inexpensive, obvious, or prescribed by law, regulation, or standard practice. Deviating from legal requirements generally requires a waiver or variance from an official (such as a fire code inspector), while it is generally imprudent to vary from standard practices such as ANSI or DIN standards without careful analysis.

12.3.9 Revised risk

If the study team is making specific recommendations, then an estimate of the risk **if all proposed controls were implemented** should be made. This "revised risk" will presumably be lower than that which currently exists.

If the risk level remains unacceptably high, more controls should be recommended. Alternatively, careful consideration should be made of methods for achieving equivalent safety or removing the hazard entirely. This last approach is an excellent opportunity for applying the methods of **inherently safer design** (see section 15.3 below) [1].

If the scope of the study is limited to identifying unacceptable risks, it is generally not possible at time of analysis to determine what degree of risk reduction has been achieved. An estimate of revised risk should still be made, but this step may need to occur after remedial measures are determined offline (as this often takes time and management/principal investigator attention and decision-making).

12.4 Example What-If? Study: Multi-axis press

In the basement of an engineering building sits a multi-axis press designed to allow researchers to test the behavior of building materials under high stress, twisting, and shaking, as in an earthquake or hurricane. The press is capable of exerting up to 100 tons vertically, with 25-ton loads applied laterally and an overturning moment added to the mix. The one-of-a-kind press helps civil engineering researchers design stronger columns, wall and roof systems, and other structural elements to better withstand extreme conditions.

The press is complex and involves very strong forces; there is the potential for catastrophic failure. In order to help assure that the press is constructed and operated safely, a What-If? Study would be an appropriate tool. Three What-If? questions are shown below, with all components from possible causes to an evaluation of the final risk. The corresponding record sheets are shown in Table 12.2; where appropriate, notes are given in the text below.

Table 12.2: What-If? study example worksheets

What-If? Worksheet

Date: 2016-03-13
Node/System section: Hydraulic actuators
Study team: Prof. Plum, Ms. White, Mr. Green, Ms. Scarlet

What-If?	Causes	Consequences	Controls	Initial risk	Recommendations/Actions (Assigned to)	Final risk
Hydraulic actuator fails?	Seal leak Piston break Dirty hydraulic fluid Insufficient hydraulic fluid	Applied force low or non-existent Unbalanced sample from low force on one axis Sample slips from mount Sample fails in unexpected mode	High actuator quality Periodic visual maintenance checks of actuator	H (6)	Periodic maintenance of actuators per manufacturer Periodic functional tests of actuators Check hydraulic fluid level before experiment Change or clean hydraulic fluid semi-monthly Enclose sample mount in shield	M/L (3/4)

Table 12.2: (continued)

What-If?	Causes	Consequences	Controls	Initial risk	Recommendations/Actions (Assigned to)	Final risk
Hydraulic line fails	Poor maintenance Sample drops on/dragged over hydraulic line Hydraulic pressure too high Abraded hydraulic line Poor-quality hydraulic line	Applied force low or non-existent Unbalanced sample from low force on one axis Sample slips from mount Sample fails in unexpected mode Hydraulic fluid sprays around laboratory Possible hydraulic flfluid fire Slip hazard from spilled hydraulic fluids	Periodic maintenance checks of hydraulic lines Researchers wear work boots with slip-resistant soles	H(6)	Raise hydraulic lines to avoid sample drags/drops Backup hydraulic pressure shutdown on hydraulic pump Write SOP for hydraulic fluid changes to prevent wrong fluid being used Periodic replacement of hydraulic lines per manufacturer recommendations Quality control on hydraulic lines Enclose sample mount in shield Use nonflammable/low-flammability hydraulic fluid Provide spill control supplies for hydraulic fluid	L(3)
Hydraulic pump leaks	Poor maintenance Failed shaft seal Dirty hydraulic fluid	Low hydraulic pressure Loss of hydraulic fluid to floor Slip hazard from spilled hydraulic fluid Hydraulic actuators fail	Periodic visual checks of pump seals Researchers wear work boots with slip-resistant soles	M(5)	Write SOP for hydraulic fluid changes to prevent wrong fluid being used Periodic shaft seal maintenance/replacement per manufacturer instructions Provide spill control supplies for hydraulic fluid	M(5) or L(3) – see text

12.4.1 Nodes

We first divide the press into **nodes** for purposes of making the study easier and more systematic. Possible nodes might include:
1. Reaction frame
2. Lateral reaction frame
3. Specimen mount and specimen
4. Hydraulic actuators
5. Loading platen
6. Control systems
7. Laboratory utilities

We will study the hydraulic actuators.

12.4.2 Team assembly

A suitable team for this analysis would include the following personnel:
1. A facilitator trained in the What-If? Method (Miss Scarlet);
2. A "scribe" to ensure that the study is fully recorded and no recommendations or risks are missed (Mr. Green);
3. The designer of the apparatus (Professor Plum); and
4. A graduate student/laboratory researcher familiar with the press' operation (or, if it were not yet built, familiar with other large presses (Ms. White).

12.4.3 What-if #1: What if a hydraulic actuator fails?

We will limit ourselves to actual failure of the actuator here, rather than asking the broader question "what if there is no hydraulic actuation?" This is a matter of preference; the second question is an entirely valid "What-If?" question.

12.4.3.1 Causes
Causes of hydraulic actuator failure are straightforward.

12.4.3.2 Consequences
If one actuator fails, there will be little or no force applied on one axis. Depending on the design of the sample mount and the nature of the sample, the test sample could slip out of the mount (falling on a researcher) or fail catastrophically, threatening the safety of the researchers operating the machine.

12.4.3.3 Controls
Currently, we rely on the quality of the actuators and periodic visual checks of the actuator system as protection against failure. This is not a particularly comprehensive set of safety checks, and so the probability of the failure is higher than it would otherwise be.

12.4.3.4 Current risk
This failure could easily lead to maiming (or possibly death) of a researcher struck by pieces of a failed sample, and given the small scope of the controls provided, it is a relatively probable event. We consider the current risk of this What-If? to be high; if we were using the system of section 3.3 above, we would give the event a severity of 4, a probability of 4, and a mitigation of perhaps 2, for an overall risk of 6. We conclude that we should consult safety professionals to identify additional mitigation; in this case, we can make recommendations for mitigating measures to bring the risk down.

12.4.3.5 Recommendations
We make several recommendations, each to specifically address causes or consequences of the event:
1. Periodic maintenance of actuators as specified by the manufacturer;
2. Periodic functional tests of actuators;
3. Checking the hydraulic fluid level before an experiment;
4. Cleaning or replacing the hydraulic fluid semi-monthly;
5. Enclosing the sample mount in a shield strong enough to withstand a failed sample, thus protecting the operator; alternatively, a shielded station could be provided for the operator.

12.4.3.6 Revised risk
Using the system established above, severity and probability remain at 4, as these are rated without consideration of added mitigation. The mitigation would change to a 4 or 5, depending on the quality of the shield provided to protect operators, leaving the risk at a 3 or 4, a medium or low level.

12.4.4 What-if #2: What if a hydraulic line fails?

12.4.4.1 Causes
Aside from the obvious "dropped sample on hydraulic line," lines might fail from abrasion (e.g., from researchers walking on them repeatedly), from excessive pressure fluctuations, or from inherent quality issues.

12.4.4.2 Consequences

Many of the consequences of a broken hydraulic line are the same as for the previous actuator failure, since a line break can cause actuators to not function. A broken hydraulic line adds hazards associated with spraying hydraulic fluid (which may be combustible) around the lab, where it could find an ignition source or lubricate a surface on which researchers could slip.

It should be noted that even though most hydraulic fluids are not technically **flammable**, many are **combustible**. Finely-divided combustible fluids dispersed in air, such as might happen when fluid is released from a failed hydraulic hose, are frequently flammable. This is because of the increased surface area available for combustion.

12.4.4.3 Controls

We assume that researchers wear slip-resistant work boots in this particular lab; such PPE provides protection against slips and falls due to spilled hydraulic fluid.

12.4.4.4 Current risk

Due to the possibility of fire, we rate the severity of a broken line at 5; the probability of a line actually breaking is somewhat lower than a mechanical failure of an actuator, so we rate it at 3. Again, the only mitigation provided is PPE and visual inspection of the apparatus, so mitigation credit is only 2. This leaves us with a high estimated risk (6), similar to the previous What-If? question.

12.4.4.5 Recommendations

A variety of recommendations are possible, given the plethora of causes and consequences of this particular scenario.

12.4.4.6 Revised risk

Note that the recommendations do not remove the possibility of a broken hydraulic line but merely reduce its probability. In some cases, they provide design features, such as elevated hydraulic lines, that reduce the probability of a broken line, but not all causes and consequences are addressed in this manner—in particular, use of the appropriate hydraulic fluid is guaranteed only by a standard operating procedure—an administrative control. We thus only increase the mitigation credit to 3, leaving the "residual risk" at 3.

12.4.5 What-if #3: What if the hydraulic pump develops a leak?

12.4.5.1 Causes

Pump leaks often arise from problems with the shaft seal. In some cases, this can be caused by internal abrasion from dirt in the fluid.

12.4.5.2 Consequences

If the hydraulic pump leaks a significant portion of its fluid, the hydraulic actuators on the apparatus will fail. Nevertheless, the catastrophic sample failure that might occur if a single actuator failed is much less likely here, since all the actuators will fail, and the sample will no longer be under stress.

We assume for the purposes of analysis that a shaft seal leak would not spray fluid around the lab in the manner a broken hydraulic line would. This should be checked against the design of the pump itself and the issue revisited if the assumption is found to be wrong. It is fine to make assumptions over the course of a safety analysis, but each assumption should be recorded and checked before the analysis is finalized.

12.4.5.3 Controls

Controls are similar to the previous scenario.

12.4.5.4 Current risk

This scenario does not maim or kill anyone, although injury is possible from a fall (severity = 3). Seal leaks are fairly common in hydraulic systems, so we rate the probability at 4. Only PPE and administrative controls are provided to prevent problems, so we take 2 for mitigation credit, revealing the current level of risk to be medium (5).

12.4.5.5 Recommendations

Since many of the causes and consequences are similar to those of the previous scenario, the corresponding recommendations are repeated.

12.4.5.6 Revised risk

Note here that we have not provided any recommendations that actually change the mitigation credit—there are no design changes that reduce or eliminate the probability or severity of the scenario. This means that, numerically, our residual risk is the same medium (5) level with which we started. Certainly, implementing the recommendations makes that medium risk more tolerable, but this is a sign that we should consider other methods for reducing risk further.

In this case, we could enter a recommendation to switch to "canned" hydraulic pumps, which do not have shaft seals to leak. We could then increase our mitigation credit to 5 and reduce the residual risk to low (3) level.

12.5 Exercise: What-if? technique

Conduct a What-If? study on an experiment you have performed or participated in previously. Consider the following:
A. What were the main risks identified?
B. Were you aware of these risks at the time you did the experiment?
C. Were those risks adequately controlled? If not, what could or should have been done to control them?

Part IV: **Practical applications of hazard control**

13 Controlling hazards in a laboratory procedure using JHA

13.1 Reproducing a procedure from the literature

Electropolishing is a metallurgical procedure sometimes called "reverse electroplating." A metal piece with a rough surface, perhaps fresh-cut, is immersed in an electrolyte, acting as anode to a pre-installed cathode. Upon application of direct current, metal is removed from the surface preferentially at the high points, leaving a smooth surface. Depending upon how current, electrolyte composition, agitation, and temperature are manipulated, as little as 0.0025 mm (0.0001 inch) can be removed by the operation [114].

In reproducing a procedure from an old (1950s-era) paper, a laboratory experienced a chemical "eruption" while mixing up the required electropolishing electrolyte. Because electropolishing uses mixtures of oxidizing acids such as perchloric acid and organic compounds such as acetic anhydride, energetic reactions (or even explosions) are not unknown; the worst electropolishing incident (in Los Angeles, CA, US in 1947), killed nearly 20 people and injured over 100. The compounds responsible were perchloric acid and acetic anhydride; to this day, it is illegal to use mixtures of these materials for electropolishing in Los Angeles.

The particular paper was quite sketchy on the procedure to be followed to mix the electrolyte; perhaps it was assumed that the reader knew how to do this as a basic laboratory operation. The electrolyte was a form of "nital," a mixture of nitric acid and ethanol, a common electropolishing electrolyte viewed as "safe." In actuality, nitric acid and alcohols react strongly: nitric acid-ethanol mixtures are only stable to 5–10% nitric acid, and nitric acid and propyl alcohol combust on contact [78]. Even when an explosion or fire does not occur, the mixing releases a great deal of heat.

A Job Hazard Analysis for the operation might look like Table 13.1.

Measuring out the chemicals involve the potential for a spill or splash, so controls are oriented toward handling those possibilities. Note that nitric acid is considerably more hazardous than ethanol, and so personal protective equipment suitable for an oxidizing acid is specified for steps involving nitric acid.

The mixing step can be expected to release energy, most likely in the form of heat of mixing; because the researchers should anticipate the possibility of an energetic chemical reaction, extra controls were added. In particular, the mixture should be made on a small scale (30 mL or so) and observed at room temperature for several hours to get a sense of its stability. The operation is specified to be conducted slowly, with cooling. It is also done in a chemical fume hood with the sash lowered, and with the extra addition of a heavy explosion shield between the operation and the operator.

Table 13.1: Job Hazard Analysis for electrolyte mixing

#	Step	Hazard(s)	Control(s)
1	Assemble materials	Drop glassware Drop chemical bottle	– Use bottle carriers for chemicals
2	Measure ethanol	Ethanol spill Ethanol splash (eye hazard)	– Conduct operation in secondary containment – Spill control kit on hand – Conduct operation in chemical fume hood with sash lowered – Lab equipped with eyewash – Personal protective equipment: lab coat, gloves, chemical splash goggles
3	Measure nitric acid	Nitric acid spill Nitric acid splash	– Conduct operation in secondary containment – Spill control kit on hand – Conduct operation in chemical fume hood with sash lowered – Lab equipped with eyewash/safety shower – Personal protective equipment: lab coat, acid apron, heavy acid-resistant gloves, chemical splash goggles, face shield
4	Add nitric acid to ethanol	Heat release Possible eruption Possible chemical reaction Nital spill Nital splash	– Add slowly – Stir during addition – Conduct operation in chemical fume hood with sash lowered – Conduct operation behind splash/explosion shield – Conduct operation with cooling (immerse beaker in pail of ice) – Try first on small scale, allow to sit at room temperature for 4 h to observe behavior – Spill control kit on hand – Lab equipped with eyewash/safety shower – Personal protective equipment: lab coat, acid apron, heavy acid-resistant gloves, chemical splash goggles, face shield

⚡ When reproducing procedures from the literature, it is **essential** to investigate both the procedure itself and your ability to carry it out. Just because the authors were able to conduct the operation without serious incident does not mean a) the procedure is actually safe, b) all essential steps are included in the paper, or c) you have adequate laboratory skills to handle the operations involved. In the laboratory experiencing the eruption, researchers did not stir or cool the mixture, simply dumping the ingredients together; naturally, after a few minutes, the chemicals mixed and the heat of reaction caused the eruption. When the composition specified by the procedure was examined further, it was found that the 20% nitric-acid-in-ethanol mixture was inherently unstable, and safety specialists estimated that the large beaker contained approximately 0.25 kg (0.5 lb) TNT-equivalent of energy.

13.2 Exercises: Using procedures taken from a research paper

Select a journal article which interests you from those in the list below. Perform a Job Hazard Analysis based on the article. Be certain to pay close attention to the Methods & Materials section, as well as (at least) skimming the article for other hazard-related information. Your instructor may specify a particular paper for you, may have a different list, or may ask you to go into the literature and select an article.

All articles listed below are available Open Access at DeGruyter.com and are licensed under the Creative Commons Attribution-NonCommercial-NoDerivatives 3.0 License.

1. Haddad, M; Soukkarieh, C; Khalafl, HE; Abbady, AQ. Purification of polyclonal IgG specific for Camelid's antibodies and their recombinant nanobodies. **Open Life Sciences.** Volume 11, Issue 1, Pages 1–9. DOI: 10.1515/biol-2016-0001, February 2016.
2. Roman, A; Popiela-Pleban, E; Migdał, P; Kruszyński, W. As, Cr, Cd, and Pb in Bee Products from a Polish Industrialized Region. **Open Chemistry.** Volume 14, Issue 1, Pages 33–36, DOI: 10.1515/chem-2016-0007, March 2016.
3. Horvatinčić, N; Sironić, A; Barešić, J; Bronić, I; Nikolov, J.; Todorović, N; Hansman, J; Krmar, M. Isotope analyses of the lake sediments in the Plitvice Lakes, Croatia. **Open Physics.** Volume 12, Issue 10, Pages 707–713, DOI: 10.2478/s11534-014-0490-7, August 2014.
4. Drzewicki, W; Ciężka, M; Jezierski, P; Jędrysek, MO. Variability of sulfur speciation in sediments from Sulejów, Turawa and Siemianówka dam reservoirs (Poland). **Open Geosciences.** Volume 7, Issue 1, DOI: 10.1515/geo-2015-0027, June 2015.
5. Zheng, M; Sun, SM; Hu, J; Zhao, Y; Yu, LJ. Preparation of Nano-Composite $Ca_2-\alpha Zn_\alpha(OH)4$ with High Thermal Storage Capacity and Improved Recovery of Stored Heat Energy. **Open Engineering.** Volume 5, Issue 1, ISSN (Online) 2391-5439, DOI: 10.1515/eng-2015-0002, November 2014.
6. Salcedo, K; Torres-Ramírez, E; Haces, I; Ayala, M. Halogenation of β-estradiol by a rationally designed mesoporous biocatalyst based on chloroperoxidase. **Biocatalysis.** Volume 1, Issue 1, DOI: 10.1515/boca-2015-0001, April 2015.

14 Evaluating risks in an experimental apparatus using What-If? technique

14.1 Case study: What-If? technique

The behavior of materials in extreme environments is of great interest for many applications. One of these extreme environments is shock: a research physiologist may wish to study the behavior of the eye under shock to better treat victims of improvised explosive devices; an artillery designer may wish to study how armor behaves under impact from a projectile; the manufacturer of a reinforced window may wish to study how its product behaves when struck by a football.

Studies of this nature can be done using a **Kolski bar** (or Hopkinson bar) apparatus [115]. A Kolski bar is essentially three metal bars connected to a gun powered by inert gas: when the valve to a gas reservoir is opened, gas rushes into the breech of the gun, pushing a bar out of the gun barrel. The bar strikes a second bar like a hammer strikes a nail, producing a shock wave that travels down the second bar. A small sample is held between the second bar and a third, and a camera or other imaging instrument is focussed upon the sample. When the shock wave reaches the sample, the camera can study the sample's behavior; a flashgun or laser illuminator serves to light the picture. The sample is generally pulverized by the shock energy. The shock wave continues on into the third bar, which leads to a damper that absorbs the remaining energy. See Figures 14.1–14.3.

Figure 14.1: Kolski bar gun breech.

Figure 14.2: Kolski bar sample containment. Flashlamp for camera is to the right of the compartment; laser illuminator to the left.

Figure 14.3: Kolski bar energy dump.

Table 14.1: What-If? study worksheets for Kolski bar

What-If?	Causes	Consequences	Controls	Initial risk	Recommendations/Actions (Assigned to)	Final risk
Inert gas pressure too high	Regulator failure Tampering with regulator Mistake in setting working pressure	Pressure safety relief valve opens Inert gas to lab atmosphere Lab atmosphere unbreathable Personnel death Burst pressure hose Personnel struck by hose	Pressure safety relief valve (protects against burst hose) Periodic hose changes (maintain strength) Standard operating procedures (specifies allowable pressure)	M(4)	Post "Possible Oxygen-Deficient Atmosphere" sign on door Install oxygen level alarm Route pressure safety relief valve output to building exhaust plenum	L(2)
Gun fired while person standing next to it	Procedural violation	Noise exposure Personnel struck by pulverized sample Personnel injury	Standard operating procedures (specify exclusion zone around apparatus while charged)	L(3)	Enclose sample strike area with shatter-resistant plastic (1 cm polycarbonate)	L(2)
Firing solenoids fail closed	Poor maintenance Solenoid beyond service life Solenoid failure	Gun remains charged	None	M(5)	Maintain solenoids per manufacturer recommendations Replace solenoids after 80% service life Test solenoids periodically Provide gas bleed port to remove gun charge (route exhaust to safe location)	L(2)
Sample shatters to shrapnel	Too high impact energy set Sample more brittle than expected	Shrapnel thrown around lab Personnel injury Personnel maiming (eyes)	Safety glasses specified while gun charged	M(5)	Enclose sample strike area with shatter-resistant plastic (1 cm polycarbonate)	L(3)

14.1 Case study: What-If? technique — 205

Table 14.1: (continued)

What-If?	Causes	Consequences	Controls	Initial risk	Recommendations/Actions (Assigned to)	Final risk
Sample is hazardous material	Sample not fully characterized before testing (e.g., sample contains asbestos) Deliberate testing of hazardous material	Laboratory contaminated with hazardous material Hazardous material exposure to personnel Personnel injury or death (depending on material)	None	H(8)	Consult safety professionals for appropriate containment, PPE, and other precautions before testing hazardous materials Do not accept inadequately-characterized samples	M(4)
Change made to gun or gas tubing	Deliberate change (new tubing material) Non-deliberate change (inadvertently replacing tubing with another brand)	Damage to gun/tubing Malfunction of gun/tubing Personnel injury/death	Informal management-of-change procedure	H(7)	Formalize management-of-change procedure Require approval of all changes by laboratory manager	M(4)
Class 4 laser illuminator used	Deliberate modification	Personnel ocular or skin laser exposure Fire	Laser safety eyewear provided	M(5)	Enroll laser in laser-control program Provide Standard Operating Procedures for laser use Remove combustible items from laser area Enclose laser area (reduces overall installation to Class 1)	L(2)
Bar shatters	Fatigue/embrittlement Flaw in bar Gas gun pressure much too high	Metal shrapnel to lab Personnel injury or death (by impalement)	None	H(7)	Limit number of shots per bar Overdesign bar X-ray bar to identify inclusions, etc. Address high gas pressure (above)	L(2)

Because of the amount of energy involved and the pulverization of the sample, there are several risks involved in this experiment. The What-If? method is suitable for study and amelioration of these risks. We can easily imagine suitable What-If? questions for the study:
1. What if the inert gas pressure is too high?
2. What if the gun is fired while someone is next to it?
3. What if the gun's firing solenoids fail closed?
4. What if the sample shatters instead of pulverizes, creating high-velocity shrapnel?
5. What if a researcher uses a sample that is a hazardous material?
6. What if someone makes a change to the gun or its gas tubing?
7. What if a Class 4 laser is used to illuminate the sample?
8. What if one of the bars shatters?

And so on. Table 14.1 provides an abbreviated What-If? analysis for the Kolski bar system.

14.2 What-If? technique study on an experimental apparatus

Select a journal article which interests you from those in the list below. Perform a What-If? Analysis based on the article. Be certain to pay close attention to the Methods & Materials section, as well as (at least) skimming the article for other hazard-related information. Your instructor may specify a particular paper for you, may have a different list, or may ask you to go into the literature and select an article.

Your objective is to produce a prioritized list of recommendations to improve safety and reduce risk.

All articles listed below are available Open Access at DeGruyter.com and are licensed under the Creative Commons Attribution-NonCommercial-NoDerivatives 3.0 License.
1. Ndlovu, SC; Chetty, N. Experimental determination of thermal turbulence effects on a propagating laser beam. **Open Physics**. Volume 13, Issue 1, DOI: 10.1515/phys-2015-0028, August 2015.
2. Changcharoen, W; Somravysin, P; Eiamsa-ard, S. Thermal and fluid flow characteristics in a tube equipped with peripherally-cut dual twisted tapes. **Open Engineering**. Volume 5, Issue 1, ISSN (Online) 2391–5439, DOI: 10.1515/eng-2015-0009, December 2014.
3. Jędrzejczyk, T; Kołaciński, Z; Koza, D; Raniszewski, G.; Szymański, L; Wiak, S. Plasma recycling of chloroorganic wastes. **Open Chemistry**. Volume 13, Issue 1, DOI: 10.1515/chem-2015-0023, December 2014.
4. Ding, K. Organic-inorganic interactions in the system of pyrrole-hematite-water at elevated temperatures and pressures. **Open Geosciences**. Volume 7, Issue 1, DOI: 10.1515/geo-2015-0050, November 2015. [NOTE: For this paper, you should perform a What-If? Analysis on the laboratory procedure as opposed to the apparatus.]

15 Designing an experiment from scratch

When first conceiving an experiment, there are many factors to be considered: the effectiveness with which the experiment can test the hypothesis; practical matters of construction; experimental accuracy and precision; cost; etc. It is **essential** to consider safety as one of these initial "up-front" issues. Treating laboratory safety considerations as exclusively later-stage work creates a very high risk that extensive redesign will be required, substantial rework and "safety add-on" costs will be incurred, and even that the experiment will need to be abandoned. Safety must be **built-in** to an experiment, not **bolted-on**.

15.1 Hazard controls are *ex post facto* solutions

The traditional hazard control process—hazard identification, risk assessment, hazard analysis, and hazard control—is systematic and effective, but it is also an **ex post facto** solution to safety issues. Controlling hazards after they appear is time-consuming, error-prone, and expensive both in terms of funds spent and rework time wasted. It is most effective to control hazards **before they appear**.

15.2 The only set factor in an experiment is the objective

The key to efficient and successful experimental design is the realization that **the experimental objective is the sole set factor in an experiment**—the only thing about an experiment that is actually fixed is the hypothesis to be tested. Everything else about the experiment is subject to debate and change.

It is very common for experiments to be performed in certain ways because "it's always been done that way," or because certain pieces of equipment (or certain materials) are easily available, etc. These decisions are often made at the level of prejudice, that is, thoughtlessly, without active consideration. This does **not** guarantee that the experiment is performed in the most scientifically useful or safest manner.

It is essential for experimentalists to question the basic assumptions about **how** the experiment is to be conducted in order to avoid increasing risk and wasting time and money on suboptimal solutions to the problem. Fortunately, a method called **Inherently Safer Design** exists to assist with this process.

15.3 Inherently safer design principles

15.3.1 The history of ISD

In 1974, at a large chemical plant in Flixborough (UK) owned by Nypro, Inc., an explosion and fire occurred that destroyed the plant, killed 28 people [116], and damaged about 1500 buildings, many in the village of Scunthorpe, about 5 km (3 miles) from the facility. The plant, which made an ingredient for the manufacture of nylon, was designed in such a manner that it had an extremely large quantity of flammable chemicals in-process at any given time.

Study of the Flixborough incident, as well as several other high-profile catastrophic accidents in the chemical industry, led chemical engineer Trevor Kletz to the conclusion that the accidents were made possible by fundamental flaws in the design process used to create the plants. Kletz published a seminal paper titled, "What you don't have, can't leak" [117]. In the paper, Kletz set forth four principles (later expanded to six by modern authors) for designing chemical processes so that **hazards do not occur** instead of **controlling hazards as they occur**. These principles came to be known as Inherently Safer Design (ISD).

An example of an "inherently unsafe design" appears in the accident at Bhopal (Madhya Pradesh, IN) in 1984. The Bhopal incident [118], which resulted in approximately 5000 casualties, involved release of approximately 42 tons of methyl isocyanate (MIC), a highly toxic chemical manufactured on-site in large batches, in a populated area. There are several competing hypotheses to explain the cause of the incident (including sabotage), but fundamentally, the cause was that the plant **had about 150 tons of highly-toxic material on-site unnecessarily**. The plant had taken multiple steps to prevent mishap with these materials, and over the years, each active hazard control had failed, leaving the plant vulnerable to the accident. An **inherently safer** way of handling MIC would have been to **use it as it was made**—minimizing the quantity of material present at any one time. **What is not present cannot be released, accidentally or otherwise**. After the incident, a sister plant in Institute, West Virginia (US) was modified (under intense public pressure) to remove its intermediate stores of MIC.

While ISD began in the chemical industry (and is on the verge of becoming mandatory for certain kinds of plant design), **ISD principles apply to biological or mechanical laboratory research just as effectively as to large-scale chemical manufacturing**. Consider the case of Preston Brown (see section 1.1.5 above). The fundamental error in the incident, aside from foolishly grinding shock-sensitive materials, was the use of **far too much** hazardous chemical—the exact same error as occurred at Bhopal. Inherently safer design would have identified the large quantity as an inherent source of risk in the experiment and indicated that the amount should have been **minimized**.

15.3.2 ISD design principles

The six modern principles of ISD are **guide words**, tools for finding inherently safer experiments and operations. They include:
- **Substitute;**
- **Minimize;**
- **Moderate;**
- **Simplify;**
- **Tolerate error; and**
- **Limit effects.**

The guide words are combined with the various properties and materials of the experiment to brainstorm ways to reduce or eliminate hazard.

For example, consider an experiment (actually proposed by a history of science class) testing the viability of an ancient Chinese weapon that used black powder to propel wooden spears and clear an area of enemy personnel; the experimenters planned to test the device in the university quadrangle. The device is a large wooden tube containing 50–100 projectiles; tied to the base of each projectile is a small bag of black powder propellant.

Applying the ISD guide words to the experiment might yield a plethora of suggestions to improve the safety of the work:
- **Substitute:** (materials) Perhaps compressed air could be used as the propellant instead of explosive black powder, which requires both a license and extensive training to handle safely.
- **Minimize:** (experimental scale) Downsizing the experiment from the original 2 m long design to more of a 10 cm "model rocket" size would reduce the hazard to personnel (although bystanders could be injured rather severely, actual death would be less likely).
- **Moderate:** (conditions) Using only enough black powder to propel the projectiles 5 m from the launcher would drastically reduce the area within which the experiment posed a hazard to people.
- **Simplify:** (equipment) Testing the device with only a few (or one) spear would prove the principle and make the experiment more controllable, while reducing the chance that bystanders or experimenters would be struck by a stray projectile.
- **Tolerate error:** (testing area) Making the test area larger by a factor of two or three than the greatest calculated range of the device would allow for errors made in design and manufacture.
- **Limit effects:** (range) If the spears were somehow tethered to limit how far they could fly, the test area required could shrink substantially.

All of these ideas remove or reduce hazards in the experiment, reducing the risk of conducting it. Note that it is still necessary to evaluate each idea for viability—ISD is fundamentally an idea-generation method, and further analysis is required to identify useful solutions from the flood of ideas. Use of inherently safer design methods does not guarantee a safe experimental design—in the case of the ancient Chinese weapon, the idea was abandoned for safety reasons—but it improves the chances of achieving a safe design and can reduce both risk and costs.

15.4 Case studies in laboratory ISD

15.4.1 Lab ISD case study: Impact testing of steel

Steel behaves differently below about −20 °C (253 K, −4°F): it becomes **brittle**. Boiler and pressure vessel regulations often require different grades of steel at low temperature, to avoid an accidental strike from a tool or similar from shattering the vessel or its associated piping. A laboratory is studying the behavior of steel under impact at low temperature using a laser interferometry technique.

A 1 m^2 square sample of steel is mounted on a flat anvil. The anvil has channels running through it so that it can be refrigerated, bringing the sample to test temperature. A 100 kg depleted uranium weight is arranged to be dropped on the steel from a height of about 3 m (10 ft). A Class 4 laser beam runs across the room, makes several elevation changes, reflects off the impact point, and is projected onto an optical table holding an interferometer.

There are quite a few places Inherently Safer Design could be applied in this experiment.

Substitute: A less powerful-laser could be used, provided it could achieve the experimental objective. It is unlikely that a Class 4 laser is necessary for simple surface interferometry. One might also substitute another dense material such as tungsten for the mildly-radioactive and moderately-toxic depleted uranium.

Moderate: The conditions under study are extremely severe, much more severe than would likely be encountered in the field. Thought should be given to using a smaller drop-weight and therefore a lower impact energy.

Simplify: The laser path is quite complex, and changing it to reduce its extent and the amount of elevation change would be desirable. Generally, a vertical component to the laser path should be avoided if at all possible, and laser beams should be kept away from eye level.

The most important Inherently Safer Design idea applicable to this experiment, though, would be **minimize!** The experiment is studying a spot no more than 5 mm diameter; there is little need to use a 1 m^2 sample. Using a small coupon of steel enables minimization of all parts of the apparatus:
- The drop-weight can be replaced by a standard Charpy test machine (see Figure 18.3).
- The laser setup can be reduced because the beam need no longer traverse a meter-square sample. The entire experiment might fit on a standard optical table.

- There is no need for the refrigerated stage. The experiment is quite rapid. Experimenters could use a nearby freezer to bring coupons to test temperature and test them immediately upon removal from the freezer.

Making these changes could easily improve the accuracy and precision of the results Using Inherently Safer Design in this instance would allow the experimenters to "step out of the box" and consider designs that both improve safety and enhance the quality of the scientific results.

15.4.2 Lab ISD case study: The "Rainbow Experiment"

The "Rainbow Experiment" is a classic general chemistry demonstration in which metallic salts are placed in a flame so that the colors of their various emissions spectra can be observed. This experiment, though, can be done in a highly-hazardous way, and it has resulted in multiple disasters in K–12 (pre-college) schools over the past several years [119].

A typical incident involving the Rainbow Experiment occurs when the flame source used is an open pool of methanol, perhaps contained in a watch glass. A small sample of several different metal salts, such as sodium chloride (for a bright yellow flame), potassium chloride (for a lilac-colored flame), and strontium chloride (which emits a red glow upon burning), are placed in watch glasses and a small sample of methanol added to each. With students gathered around to observe, the methanol is ignited and the "rainbow" of colors observed.

The incidents that have given this experiment a bad reputation have all occurred when the methanol is nearly exhausted and the instructor unwisely attempts to add more methanol to prolong the demonstration. Since methanol is a very light molecule, it is quite volatile—pouring more liquid methanol out of the container necessarily pours a large amount of vapor at the same time. In several instances, the methanol vapor has ignited; if the class is particularly unlucky, the burst of flame from the ignited vapor will vaporize more methanol from the pool or the bottle, which feeds the explosion, creating a small-scale BLEVE (**boiling liquid-expanding vapor explosion**). Students and teachers have both been severely injured.

Recently, Jillian Meri Emerson, at the University of California, Davis (US) chemistry department, turned her attention to finding a safer way to do the Rainbow Experiment. In a classic example of Inherently Safer Design, Emerson returned to the experimental objective—to heat metallic salts to flame temperature in order to observe their emissions spectra. Methanol is not necessary to achieve this objective—other sources of flame may be **substituted** for the burning pool of volatile methanol. Bunsen or Fischer burners are the suggested flame source, since they are readily available in chemistry laboratories and lecture halls. The author has also done this demonstration using a handheld soldering torch.

The switch to a gas torch or burner as flame source necessitates changes in the demonstration procedure. In Emerson's altered procedure, long wooden sticks are soaked in saturated solutions of the salts until dry; the sticks are then held in the edge of the flame and the flame color observed. The author has also used long cotton applicators dipped in saturated salt solution and allowed to dry.

This revision of the proven-dangerous methanol-based Rainbow Experiment uses several principles of Inherently Safer Design:

– **Substitute.** The experiment uses an entirely different flame source that lacks the potential for catastrophic BLEVEs from imprudently adding more fuel.
– **Simplify.** The University of California version of the experiment removes the need for handling liquid fuels entirely.
– **Limit effects.** The worst-case accident from this demonstration is probably the demonstrator setting himself or herself on fire by getting clothing in the flame. While undesirable and potentially lethal, it is a far more limited consequence than burning the entire class with a fireball.

15.5 Exercise: Inherently safer design of a hazardous experiment

Using a paper in your field, do the following:
1. **Identify all the hazards** you can locate in the experiment.
2. **Brainstorm as many ways as possible** to make the experiment inherently safer.
3. **Eliminate any ideas** which do not act **inherently** upon the system, that is, ideas which "patch up" safety problems that already exist instead of working to reduce or eliminate them.
4. **Rate each idea** (High, Medium, or Low will suffice) according to its probable cost and its likely ease of implementation. (Use your best judgment.)
5. **Identify those ideas that you would implement.** Justify your decisions, those to implement an idea or those to discard it.

Part V: **Appendices**

16 Laboratory safety checklists (abbreviated)

Table 16.1 is a checklist of common hazard types that the author uses routinely to identify the major hazard classes present in a laboratory, an experiment, or a research program.

Table 16.1: Laboratory hazard identification checklist

Hazard	✓	Details
Chemical hazards		
Toxic chemicals		
Acute toxins		
Chronic toxins		
Carcinogens/teratogens/mutagens		
Flammables/combustibles		
Flammable liquids		
Flash point ≤ 100°F		
Flash point > 100°F		
Flammable solids		
Flammable gases		
Pyrophoric materials		
Corrosives/reactives		
Acids		
Bases		
Oxidizing/reducing agents		
Fluorine/fluorinating agents		
High-pressure/liquid oxygen		
Peroxides/peroxide-formers/azides		
Water-reactive		
Nanomaterials		
Biological hazards		
Organisms		
Laboratory animals		
Agricultural pathogen		
Human pathogen		
Human cell line (including cancer)		

Table 16.1: (continued)

Hazard	✓	Details
Recombinant/synthetic DNA		
Tissues		
Human (including cadaver or cadaver-derived)		
Primate		
Animal		
Biotoxin/Select Agent		
Radiation hazards		
Ionizing		
Hard UV, X-ray, γ-ray		
Charged particles (α, β)		
Neutron		
Non-ionizing		
Soft UV, visible, IR, microwave, radiofrequency		
Lasers		
Class 1/1M, 2/2M, 3R		
Class 3B		
Class 4		
Physical hazards		
Compressed gases/pressurized equipment		
Vacuum		
Hydraulics/pneumatics		
Electricity		
High voltage (>400 V)		
Low voltage		
Magnetics/high-current apparatus		
Stored energy (large capacitors/inductors)		
Temperature		
Cryogens		
Low temperature (<273 K)		
High temperature (>325 K)		
Sparks/hot work/open flames		

Table 16.1: (continued)

Hazard	✓	Details
Machinery		
Moving parts/pinch point/shear point		
Sharp edges/puncturing parts		
Compression/rolling object		
High rotational energy		
Stored mechanical energy (e.g., springs)		
Overhead hazards (including cranes)		
Trips/fall/slip		
Sharps/needles/glassware		
Noise/vibration		

Table 16.2 is an abbreviated laboratory hazard checklist encompassing a sampling of the most important precautions to be taken. A more comprehensive list may be found in the appendices of **Prudent Practices in the Laboratory** [120].

Table 16.2: Laboratory safety checklist

✓	Question
	General safety
	Laboratory housekeeping
	Floors clear of unnecessary equipment, spill residue, parts, cords/tubing, etc.
	Benches clear of unnecessary equipment and materials
	Adequate working bench space for each occupant
	Adequate chemical fume hood/biological safety cabinet working space for each lab occupant (2 m/6 ft typical for many disciplines)
	Chemical fume hood/biological safety cabinet clear of all equipment and materials not in use
	Personal protective equipment
	Appropriate PPE provided
	PPE in acceptable condition/maintained properly/replaced when compromised
	PPE readily available
	PPE users trained in appropriate use (e.g., proper methods for removing chemically-contaminated gloves)
	Hand-washing supplies provided in laboratory

Table 16.2: (continued)

✓	Question
	PPE selected for the task by lab manager or researchers trained in PPE selection
	Laboratory culture
	Food/drink evidence absent from lab
	Lower-level researchers are comfortable bringing up safety concerns
	Safety concerns are immediately addressed
	Safety is not viewed as "someone else's problem," even when it involves someone else's work or behavior
	Laboratory personnel regularly plan for/practice for/prepare for foreseeable emergencies
	Lab construction
	Exit doors clear
	Paths to exit doors clear and at least 1 m wide
	Safety showers/eyewashes within 10 seconds unobstructed travel from chemical/biological/radiological eye hazard use areas
	Adequate storage provided for all materials (not on bench or in hood or on floor)
	Physical hazards
	Sharps
	Sharps stored safely when not in use
	Contaminated sharps disposed immediately in sharps container
	Sharps containers provided
	Sharps containers filled no more than 75% full before sealing/disposal
	Mechanical hazards
	Machines evaluated for mechanical hazards
	Guards or interlocks provided where appropriate on mechanical equipment
	Powered equipment provided with convenient emergency shutdown
	Mechanical equipment regularly maintained
	Fire hazards
	No accumulation of flammable/combustible materials
	Flammable materials above local quantity limits stored in appropriate storage
	Flammables/combustibles kept away from heat sources (e.g., furnaces, burners)
	Precautions followed for hot work operations (e.g., clearing CFH of flammables/combustibles before torch/heat gun use)

Table 16.2: (continued)

✓	Question
	Cryogens
	Adequate means provided to vent boiloff vapor
	Adequate laboratory ventilation appropriate to volume of cryogen in use
	PPE for cryogens provided, in good condition, in use, and users trained
	Cryogen users trained in appropriate first aid
	Pressure and vacuum
	Pressure/vacuum safety relief (or engineered pressure-resistant design) provided and design/design basis documented
	Pressure safety relief routed to safe location
	Pressure-containing system designed to withstand at least $3\times$ maximum expected emergency pressure
	Regulators produce appropriate pressure ranges
	Flow control provided on compressed gas systems
	Pressure-containing systems inspected and maintained regularly (by legal/insurance authorities where required)
	Electricity and magnetism
	Extension cords used only for temporary installations
	Extension cords and other wiring protected against tripping (taped down or wiring bridge supplied)
	Extension cords, outlet multipliers, power strips, and surge suppressors loaded to 80% maximum capacity
	Extension cords, power strips, and surge suppressors not "daisy chained" (connected serially)
	Surge suppressors used only for low-current/non-motorized equipment
	Grounding provided to local electrical code
	Equipment evaluated for special grounding requirements (e.g., radiofrequency grounding, high earth resistance/ground loop materials)
	Ground-fault circuit interrupters (GFCIs) provided on all laboratory outlets excepting where absolutely necessary (e.g., high current inrush leads to false trips)
	GFCIs tested periodically and replaced on failure
	Permanent electrical wiring and modifications to building electrical systems performed by qualified electrician
	No "cheater plugs" used to adapt three-prong grounded plugs to 2-prong ungrounded outlets

Table 16.2: (continued)

✓	Question
	High voltage (>400 V) design, construction, and operation conducted only by qualified and trained personnel
	Lockout/tagout procedures in place and followed for maintenance and service procedures
	Special attention paid in lockout/tagout procedures to equipment capable of storing electrical/magnetic energy
	Strong electric/magnetic fields do not exceed allowable legal limits
	CPR/AED training available to personnel working in areas with significant electrical risks
	Robotics
	Robots evaluated for mechanical hazards, particularly pinch/shear points or "clobbering" opportunities
	Pinch and shear points clearly labeled
	SOPs provided for startup, shutdown, and emergency stop of dangerous robots
	Exclusion zones established and respected for robot operations
	Robots with recognized mechanical hazards equipped with emergency shutdown and keyed arming switches
	Chemical hazards
	Chemical waste
	Chemical waste stored in secondary containment
	Chemical waste clearly labeled with contents
	Incompatible chemicals separated in discrete secondary containments
	Incompatible chemicals not mixed in chemical waste containers
	Chemical waste compatible with container materials of construction
	Chemical waste container contents accurately tracked and reported to disposal personnel
	Distinct chemical waste storage area established
	Chemical waste does not accumulate in lab (maximum 90 days at author's institution)
	Flammables handling
	Flammable materials stored in flammables cabinets or flammables-rated refrigerator/freezers when not in use
	Flammable materials removed from storage/used only in absence of ignition sources
	Sorbent materials (**not** paper towels or the like) provided for spill control

Table 16.2: (continued)

✓	Question
	Flammable materials handled only in sealed apparatus or in chemical fume hood (or similar well-ventilated area)
	Large, conductive containers of flammable materials grounded and bonded before dispensing to avoid static buildup
	Fire extinguishers provided to local requirements
	Researchers trained to either evacuate upon encountering fire or extinguish safely when appropriate
	Chemical storage
	Highly toxic materials stored in secure storage (e.g., locked cabinet)
	Incompatible chemicals stored separately to prevent mixing in case of impact accident
	Chemicals stored by compatibility and chemical nature rather than alphabetically, by sample number, by carbon number, or other system
	Acids separated from bases (in separate cabinets if local requirement, in separate secondary containments at minimum)
	Flammables separated from corrosives (in separate cabinets; corrosive flammable materials such as acetic acid should be stored with flammables)
	Oxidizing acids (perchloric, nitric, concentrated sulfuric, etc.) stored in separate secondary containments (and separate from each other)
	Chemicals stored in secondary containment to prevent mixing upon leak or impact
	Secondary containment materials compatible with chemicals being stored
	Chemicals stored in latched cabinets or lipped shelves in seismically-sensitive areas
	Biological hazards
	Refrigerators/freezers
	Seals in good condition
	Refrigerator/freezer maintaining set temperature
	Contents listed in readily-available inventory
	Inventory updated as contents added/withdrawn
	Freezers defrosted regularly
	Cryostat/liquid nitrogen freezers stored in ventilated area
	High-value or pathogenic materials refrigerators/freezers equipped with high temperature alarm
	Biological safety cabinets
	If UV lamp used, bulb replaced on manufacturer's recommended schedule

Table 16.2: (continued)

✓	Question
	If UV lamp used, biological safety professional has verified that UV kills agents of interest in the cabinet
	Biological safety cabinet sterilized at end of day
	Radiation Hazards
	Ionizing radiation
	Regular wipe tests performed and properly documented
	Radioactive waste disposed and removed from area as soon as practical (and no later than required)
	Work performed as specified by Radiation Safety Officer
	Standard Operating Procedures written, reviewed by principal investigator & Radiation Safety Officer, researchers trained, and SOPs followed
	Procedures/experimental campaigns designed for minimum radiation exposure and radioisotope use
	Non-ionizing radiation
	Laser safety eyewear appropriate for use
	Laser safety eyewear in good condition
	Laser safety eyewear provided for all persons entering Nominal Hazard Zone
	Laser beam path terminated by non-reflective, non-combustible beam blocks
	Laser beam confined to horizontal plane to greatest extent possible
	Vertical laser beam paths enclosed in beam tubes if possible
	Nominal Hazard Zone confined to smallest extent possible
	If Nominal Hazard Zone might extend beyond the lab, entryway controls/door interlocks provided
	Laser signs provided both for laser presence and active beam
	Laser Standard Operating Procedures and training in place for all users, including emergency procedures

17 Checklist reviews for common laboratory operations

This chapter contains checklists to be used to verify common safety precautions and avoid misconfigurations or other errors in sample laboratory operations. While the checklists are non-exhaustive, they are excellent examples of the use of the Checklist technique for hazard identification, analysis, and control.

Not all questions are applicable to each possible use case, so "not applicable" is an acceptable answer to the checklist questions in some cases.

17.1 Delivering gas from a compressed gas cylinder

Many pieces of laboratory equipment rely on a flow of low-pressure gas—chromatography, optical equipment such as spectrometers, plasma etching machines, etc. The author often finds odd and dangerous gas piping setups delivering gas from high-pressure cylinders. See Figures 17.1 (where a regulator designed to produce up to 200 psig is being fed into plastic tubing with a burst pressure of around 30–40 psi) and 17.2 (which shows a step-down regulator arrangement that provides both pressure and flow control for low pressures and flow rates).

Figure 17.1: Improper setup of compressed gas regulator for low-pressure gas delivery.

DOI: 10.1515/9783110444438-018

Option 1: Use low-pressure regulator

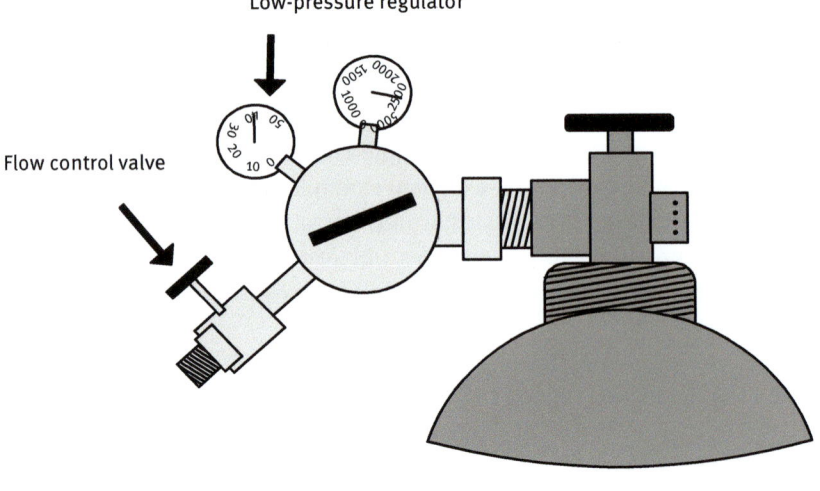

Option 2: Use secondary step-down regulator

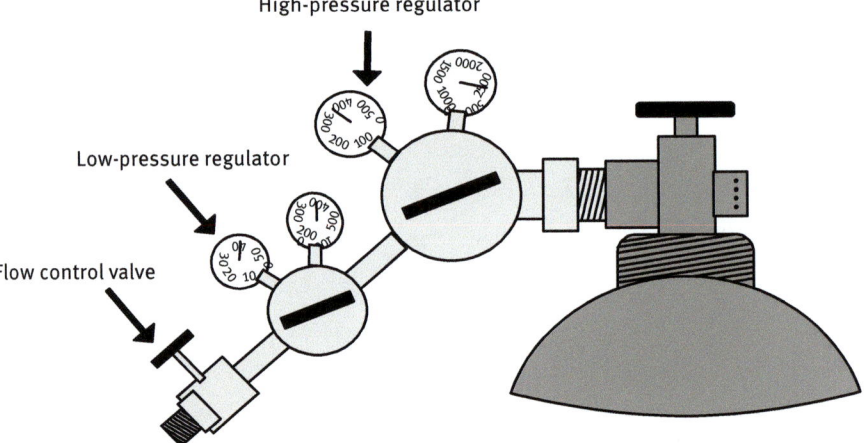

Figure 17.2: Proper setup of compressed gas regulator for low-pressure gas delivery.

Table 17.1 shows a checklist-form hazard review for delivering gas from a compressed gas cylinder.

Table 17.1: Gas delivery checklist

✓	Question
	Cylinder setup
	Is the cylinder full enough to deliver the required volume of gas?
	Is a transport cart used to move the cylinder from storage to the use location?
	Is the cylinder chained to the transport cart during transport?
	Is a wall-, bench-, or floor-mounted cylinder support available at point of use?
	Is the cylinder properly restrained? (With a chain or belt, above the midline and significantly below the neck)
	Is the cylinder cap mounted when the cylinder is in transport or not being used?
	Is the regulator appropriate for the gas and the target delivery pressure? (See Figure 17.2 for a step-down technique to deliver low-pressure gas from a high-pressure regulator.)
	Are regulator connections undamaged (no crossthreading, no scratches on sealing surfaces of CGA or DIN fittings)?
	If the regulator requires an O-ring in the seal, is a fresh one available?
	Is the delivery tubing intact and not degraded?
	Is the delivery tubing appropriate for the pressure to which it will be subjected?
	Does the regulator have the appropriate CGA or DIN fitting to connect to the cylinder. (NOTE: never change fittings on a regulator or use adaptors without consulting the manufacturer!)
	Is Teflon tape or "pipe dope" used only on fittings where the threads provide the sealing surface (e.g., NPT fittings, NOT most CGA and DIN fittings)?
	Is Teflon tape or "pipe dope" chemically compatible with the gas to be used?
	If cylinder contents are flammable or oxidizing (including high-pressure air), is a non-sparking wrench available to connect regulators and make up fittings?
	Cylinder and pressure system design and use
	Are the regulator and gas system vented to a safe location (e.g., not into the lab for hazardous gases)?
	Can the system withstand full worst-case cylinder pressure or is appropriate pressure safety relief provided?
	Is any pressure safety relief device vented to a safe location (e.g., not into the lab)?
	Is the cylinder hand valve open only enough to provide the required gas flow?
	Is a means of flow control provided?
	Are system materials (including seals) chemically compatible with the gas to be used?

Table 17.1: (continued)

✓	Question
	Cylinder breakdown and shutdown
	Is the cylinder hand valve **OFF** and fully seated?
	Has pressure been purged from the piping between cylinder and regulator?
	Is sealed storage available for the regulator once it has been dismounted (to prevent damage from dust, chemical vapors, and other mistreatment)?
	Is a sealing cap available to prevent ingress of air, dust, and moisture into the instrument or system when the regulator is not attached?
	Has the cylinder cap been reattached before the cylinder is transported back to storage?
	Is a transport cart used to move the cylinder from the use location to storage?
	Is the cylinder chained to the transport cart during transport?
	Is sufficient safe storage available at the storage location?

NOTE: This checklist is not exhaustive. The reader may wish to consult references such as the **CGA Handbook of Compressed Gases** [121] for further detail. This reference contains over 700 pages of advice in appropriate practices for gas handling.

17.2 Flame-sealing a glass tube with an oxyacetylene torch

In certain types of laboratory, it is a common operation to seal samples inside glass vessels of some sort. Glass **ampoules**, vials with a neck that can be easily scored and broken, are typical, but some laboratories prefer thin-walled NMR tubes for sample storage and transport. The sample is transferred into the tube using a nitrogen or argon manifold to protect air- or moisture-sensitive chemicals and condensed by immersing the tube in liquid nitrogen. The tube can then be sealed with a small torch (see Figure 17.3).

Table 17.2 shows a checklist for preparing to seal a glass tube.

Table 17.2: NMR tube sealing checklist

✓	Question
	Flammable materials
	Have **all** combustible materials been removed from the fume hood?
	Have **all** flammable solvents been removed from the fume hood?
	Are **all** bottles of volatile flammable solvents in the lab securely capped (other than those in closed fume hoods)?

Table 17.2: (continued)

✓	Question
	Is there **no** solvent still in the fume hood to be used for flamework?
	Are all other solvent stills in the lab in **fully closed** fume hoods?
	Emergency preparedness
	Are all other lab occupants aware that flamework is about to be done?
	Is there a clear escape plan for the lab?
	Is the flamework **not** being done in a location where a fire would block egress from the lab?
	Are all laboratory occupants trained in emergency procedures?
	If using a fire extinguisher is part of the emergency procedure, are the appropriate personnel trained in use of an extinguisher?
	If using a fire extinguisher is part of the emergency procedure, is the extinguisher present and in service?
	Assembling materials/setup
	Are all materials needed for the operation ready and within reach in the fume hood?
	If liquid nitrogen is to be used, is a **fresh**, suitably-sized Dewar of LN available?
	Is a stand and clamp available and in-place for the operation?
	Is the torch in working order?
	Is the torch fueled (with liquid fuel) or is it connected to fuel and (optionally) oxygen cylinders?
	Is the torch designed for the fuels used?
	If gas cylinders are used, are they securely fastened to a stationary object or welding cart?
	If gas cylinders are used, are they equipped with appropriate regulators?
	If gas cylinders are used, are the fuel and (if used) oxygen lines suitable for the gases they will contain and in good condition?
	Is the NMR tube mounted so that the lower half can be immersed in LN?
	Is the NMR tube mounted to allow enough space above the Dewar flask (if used) to manipulate the tube without getting one's fingers too close to the flame?
	Safety preparedness
	Are the experimenter and any bystanders wearing appropriate safety eyewear?
	(Optionally) are the experimenter and any bystanders wearing flame-retardant lab coats?

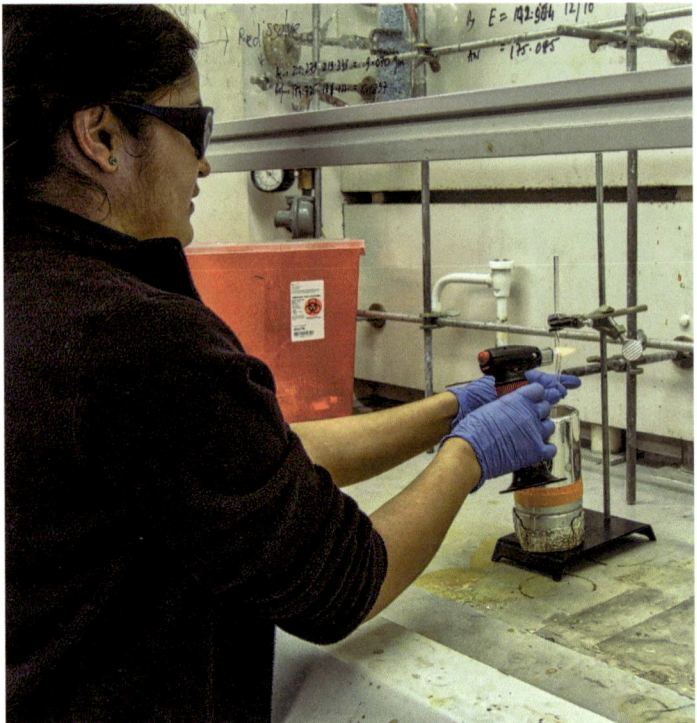

Figure 17.3: Dr Savita Sharma demonstrates appropriate techniques for NMR tube sealing with a torch.

17.3 Using a biological safety cabinet

The biological safety cabinet (BSC) is one of the most common pieces of large laboratory equipment in biological labs. It is also commonly misused. Improper use of a BSC can contaminate the user, the lab, other lab occupants, and even cause contamination threats to the public; it can also ruin experimental results. See Figure 8.8.

Table 17.3 is a brief checklist of BSC setup and use items.

Table 17.3: Biological safety cabinet setup and use checklist

✓	Question
	BSC setup
	Is the BSC **empty** of unnecessary materials and equipment?
	Has the BSC been **sterilized** or **sanitized** before use?

Table 17.3: (continued)

✓	Question
	Are the BSC fan and work light **on**?
	Is any sanitation light (UV light) **off** before use?
	Have all necessary materials (supplies, cultures, etc.) been staged on the designated "clean side" of the BSC?
	Are the baffles at front and rear of the BSC clear?
	Has the BSC been checked or certified for proper operation?
	Are air filters not in need of replacement?
	BSC operation
	Has the BSC run 3–5 minutes without interference to allow entrained particles to be filtered out?
	Does the user wait 1–2 minutes after inserting hands to allow entrained particles to be filtered out?
	Does the user **not** rest his/her arms on the front air baffles/grille?
	Is work conducted slowly to avoid bringing entrained particles into the BSC?
	Is work performed moving strictly from the "clean" to the "dirty" side of the BSC to avoid cross-contamination of samples?
	Are toxic or flammable materials used **only** in very small quantities?
	Are toxic or flammable materials used **only** at the rear of the BSC (to minimize leakage of contaminated air from the BSC)?
	Is the BSC sash lowered (if possible) if the cabinet is left alone for more than a few minutes?
	BSC shutdown
	Are all materials sanitized or properly disposed of as biohazardous upon removal from the BSC?
	Is the BSC fully sanitized once empty?
	Is the BSC sash left in its lowest position when the cabinet is not in use?
	Is the UV sanitation lamp left lit when the cabinet is not in use if lab protocol dictates it?
	Are the BSC fan and work light left off when the cabinet is not in use if lab protocol dictates it?

18 Writing experimental protocols and Standard Operating Procedures

Precise written descriptions of an operation are critical tools in many situations—laboratory experimentation included. Part lab record, part instruction manual, part training tool, and part specification, they standardize the way that operation is conducted, increasing accuracy, precision, and repeatability. For this reason, such descriptions are often called **Standard Operating Procedures (SOPs)**, and they are common devices across a variety of industries. In a laboratory context, the author finds that the term **experimental protocol** resonates better, because safety is better thought of as an integral part of the experiment instead of something "added on" later.

18.1 Types of "SOP"

SOPs are normally useful for three purposes:
- Defining how a particular experimental procedure is to be conducted (which we will call a **procedural SOP**);
- Setting rules for handling, disposal, etc. of hazardous materials, whether chemical, biological, or radioactive (a **material SOP**); and
- Explaining how to use a particular piece of laboratory equipment, such as a gas chromatograph or material tensile tester (an **equipment SOP**).

18.2 General advice on writing protocols

It is essential to be very clear about the purpose for which a procedure is prepared, and the operations for which protocols are written should be chosen carefully. Before beginning, examine exactly **how** an instrument is to be used, **what** experiments are to be conducted, and **which** materials are to be used.

It is not uncommon to write procedures for experiments, equipment, or materials that are already regularly conducted or used. In this situation, close consultation with the personnel who perform the procedure is critical. It may be appropriate to convene a meeting of all laboratory personnel who use or will be using the SOPs to discuss how it is done. Oftentimes, such a meeting will bring to light differences in the manner in which different researchers perform the task, accidents or close-calls that have almost faded from institutional memory, and tips or cautions that more experienced personnel have learned from hard experience.

SOPs usually contain three types of information: the actual steps to be performed, notes and explanations related to those steps, and optional procedural branches that cover recovery from abnormal conditions and such. **It is essential that the user be able to quickly locate the steps he or she is to perform**, and so the first type of information noted above should be physically separate from the notes and secondary steps.

Note the following other practices for writing comprehensible experimental protocols:
- Use an active, imperative writing style, that is, express each step or practice as a positive command to the user, in the order it is to be performed;
- Do not be afraid to leave empty space (called "white space" in publishing) on the page—it is an essential organizational tool and aid for locating the important information;
- Avoid fancy formatting such as excessive font changes;
- Use **boldface** or italics for emphasis, but sparingly.

For procedural and equipment SOPs, the author favors a two-column format in which the first column contains only actions to be performed, in order, while the second provides notes, secondary or emergency procedural steps, etc. Outlines or bullet-point lists often suffice for material SOPs. Procedures should **never** be written in paragraph form, instructions interspersed with notes and secondary steps. Consider the difference between the following brief procedures for sanitizing a BSC:

> The biological safety cabinet (BSC) requires sanitization before use. 70% ethanol should be used as a sanitizer for the recombinant **E. coli** samples. All surfaces should be sanitized using sanitizer sprayed on a laboratory wipe; do not spray ethanol in the BSC because an explosion could occur. In use, hands should also be sprayed (**outside** the BSC) with 70% ethanol and then inserted into the BSC.

Table 18.1: Biological safety cabinet sanitization procedure

	Step	Notes
1	PREPARE 70% ethanol in a spray bottle	– Ethanol is stored in the flammables cabinet.
2	PUT ON nitrile gloves	
3	SPRAY gloves with 70% ethanol and rub gently	– Allow gloves to air-dry before proceeding. – Avoid breathing ethanol vapor. – DANGER! EXPLOSION HAZARD! Spray OUTSIDE the BSC to avoid pulling flammable materials into the blowers and causing an explosion.

Table 18.1: (continued)

	Step	Notes
4	SPRAY laboratory wipe with 70% ethanol	– Use sufficient ethanol to barely saturate the wipe without dripping. – DANGER! EXPLOSION HAZARD! Spray OUTSIDE the BSC to avoid pulling flammable materials into the blowers and causing an explosion.
5	WIPE down interior surfaces of BSC	– Pause 60 seconds after inserting hands through air curtain to allow BSC to clear entrained particles – Use additional wipes if necessary. – Place used wipes in the waste bag in the BSC.
6	WIPE down equipment and supplies in the BSC	– Supplies that cannot be sanitized (e.g., boxes of lab wipes) should be regarded as contaminated and disposed.
7	TAKE OFF gloves	– Use appropriate sterile technique for removing gloves. – Dispose gloves in the biohazard bag outside the BSC.

Another formatting option suitable to SOPs for equipment is the format in Figure 18.1. This procedure, for launching of a Titan II nuclear missile (the Titan II system was decommissioned in the 1980's), uses the first line of each step to convey the action to be performed, in a sort of checklist format, providing the condition under which the step is successful (e.g., "OPERATE INITIATE [lamp on the launch console].........lit" to indicate that the weapon launch sequence has begun). Notes, cautions, and recovery from abnormal conditions follow, in paragraph form.

18.3 Writing protocols for hazardous materials handling

"Standard operating procedures" for hazardous materials handling reduce, for the most part, to lists of "do"s and "don't"s associated with the material, with sub-procedures for actual tasks that must be performed, such as waste disposal or spill cleanup/decontamination.

Tables 18.2 is a lightly-edited excerpt from the hazardous materials standard operating procedures for use of hazardous chemicals in use by a major research institution.

18.3 Writing protocols for hazardous materials handling — 233

T.O. 21M-LGM25C-1

STEP	
	NOTE *If flight simulation test is in progress, press MASTER RESET pushbutton on MGACG to terminate test. *Do not stop launch sequence after key turn except to react to indications that would affect successful completion of launch. *Do not perform Ordnance Circuit Test. **CAUTION** *Do not launch with DEFLECTOR HIGH LEVEL lighted due to high water level. *Do not launch when atmospheric conditions exceed T.O. 21M-LGM25C-1-2 limits.
1. MCCC	SURFACE WARNING CONTROL . . . . Lighted red If indicator is not lighted red, press pushbutton required number of times to cause indicator to light red.
2. MCCC DMCCC	Launch keys . . . . Inserted Remove security seals and insert keys into switches. **WARNING** Do not continue checklist unless sortie is executed and an EWO commit time is available for launch. If BVL is unlocked, do not reconfigure unless directed by Battle Staff.
3. BMAT	Circuit breaker 103 on . . . . Set For EWO launch, if circuit breaker trips off, attempt to reset. If circuit breaker does not reset, continue checklist.

Figure 3-1. Launch Checklist (Sheet 1 of 3)

Change 4 3-9

Figure 18.1: U.S. Air Force Technical Order 21M-LGM25C-1, "Launch Checklist" [for Titan II Missile System]. Figure courtesy Titan Missile Museum, Titan Missile Historic Landmark, c. 1985 (best available copy).

T.O. 21M-LGM25C-1

STEP	
4. BMAT MFT	BVLC - Operate Code Word . . . . Entered NOTE * If OPERATE OK on the BVLC does not light in step 5, verify that the proper code is inserted. * If proper code is not inserted, press OPERATE INITIATE to not lighted, enter proper code, and reaccomplish step 5. * If OPERATE OK does not light, refer to figure 2-11 of T.O. 21M-LGM25C-121-1 for EWO launch. Notify command post of type deviation, ETOR and intent to perform TCCPS. Post ETOR to EWO document. For peacetime launch refer to T.O. 21M-LGM25C-2-26.
5.	BVLC - OPERATE switch . . . . Pressed OPERATE INITIATE . . . . Lighted OPERATE OK . . . . Lighted **WARNING** Insure key turn time is in accordance with current EWO directives.
6. MCCC DMCCC	Launch keys, at commit time . . . . Turned, held Simultaneously (within 2 seconds) turn keys for 5 seconds or until sequence starts. EWO LAUNCH: If sequence does not start after coordinated key turn, refer to figure 2-1 of T.O. 21M-LGM25C-121-1. Notify command post of type deviation, ETOR, and intent to perform TCCPS. Post ETOR to EWO documents. PEACETIME LAUNCH: If abnormal indications occur prior to step 8, report to Launch Director and proceed as directed.

Figure 3-1. Launch Checklist (Sheet 2 of 3)

3-10   Change 13

Figure 18.1: (continued)

18.3 Writing protocols for hazardous materials handling — 235

T.O. 21 - LGM25C-1

STEP	
7. MCCC	LAUNCH ENABLE..Lighted NOTE *If launch sequence does not continue to liftoff, refer to HOLD or ABORT checklist, as applicable. *If LCCFC Launch Sequence Go indicators stop for 35-40 seconds with no hold indicator, refer to HOLD checklist.
8.	BATTERIES ACTIVATED................................. Lighted
9.	APS POWER.. Lighted
10.	S ILO SOFT.. Lighted
11.	GUIDANCE GO... Lighted
12.	FIRE ENGINE..Lighted
13.	LIFT-OFF...Lighted
	NOTE *Disregard the LCCFC ABORT indication, if a LIFT-OFF indication is present at this time. *Immediately perform Post Launch checklist.

Figure 3-1. Launch Checklist (Sheet 3 of 3)

Change 2   3-11/(3-12 blank)

Figure 18.1: (continued)

Table 18.2: Excerpt from SOP for use of corrosive materials

Low Point University
Policy EHS42: Standard Operating Procedures
for Chemical Use (By Hazard Class) in Laboratories
Revision 4.2 (2015-09-20)
Principal author: Winnie T. Pooh

Corrosive chemicals are substances that cause visible destruction or permanent changes in human skin tissue at the site of contact, or are highly corrosive to steel. The major classes of corrosives include strong acids, bases (alkalis), and dehydrating agents.

Decontamination procedures

A. Personnel

Immediately flush contaminated area with copious amounts of water after contact with corrosive materials. Remove any jewelry to facilitate removal of chemicals. If a delayed response is noted, seek immediate medical attention. Be prepared to detail what chemicals were involved.

If the exposure involves hydrofluoric acid (HF), seek immediate medical attention. Laboratories using HF must have calcium gluconate on hand and begin immediate first aid treatment.

B. Area

Decontamination procedures vary depending on the material being handled. The corrosivity of some materials can be neutralized with other reagents. Special neutralizing agents must be on hand to decontaminate areas.

Eye protection

Eye protection must be worn at all times when handling corrosive materials. Ordinary (street) prescription glasses do not provide adequate protection. Adequate chemical splash goggles must meet the requirements of the Practice for Occupational and Educational Eye and Face protection (ANSI Z87.1-1989). It is recommended that face shields be worn when a splash potential exists with corrosive materials.

Eye protection

Where the eyes or body may be exposed to corrosive chemicals, suitable facilities for quick drenching or flushing of the eyes and body must be provided within 10 seconds unobstructed travel from the work area for immediate emergency use.

Fume hoods

Manipulation of corrosive substances must be carried out in a fume hood if corrosive vapor production is anticipated.

Gloves

Appropriate gloves must be worn when handling corrosive materials.

Hazard assessment

Hazard assessment must include instruction on proper use and handling, spill control, and splash protection.

Safety shielding

Safety shielding is required any time there is a risk of explosion, splash hazard or a highly exothermic reaction. All manipulations of corrosive materials which pose a splash risk must occur in a chemical fume hood with the sash in the lowest feasible position. Portable shields, which provide protection to all laboratory occupants, are acceptable. Uses with potential for explosion must consult EHS for assistance in obtaining appropriate explosion shielding; chemical fume hoods do not offer sufficient protection.

Safety shower

A safety or drench shower must be available within 10 seconds uninterrupted travel from any area corrosive material is used.

Table 18.2: (continued)

Signs and labels
All corrosive chemical containers must be clearly labeled with the correct English chemical name. Handwritten labels are acceptable; abbreviations, codes, chemical formulae and structural formulae are not acceptable.

Special storage
Segregate the various types of corrosives. Separate acids and bases. Liquids and solids must also be separated. Specially designed corrosion resistant cabinets must be used for the storage of large quantities of corrosive materials. Store corrosives on plastic trays. Do not store corrosive materials on high cabinets or shelves.

Special ventilation
Manipulation of corrosive materials outside of a chemical fume hood may require special ventilation controls in order to minimize exposure to the material. Chemical fume hoods provide the best protection against exposure to corrosive materials in the laboratory and are the preferred ventilation control device.

Spill response
Anticipate spills by having the appropriate clean up equipment on hand. The appropriate clean up supplies can be determined by consulting the Safety Data Sheet prior to the use of the corrosive chemical.

Waste disposal
Most corrosive materials are hazardous wastes. All waste corrosives must be disposed through the University's hazardous waste program coordinated by Environment, Health & Safety. Questions regarding waste disposal should be directed to EHS. Some neutralized corrosive wastes may be drain-disposed, but special conditions apply—consult EHS for instructions.

They are written in a paragraph format:

Safety shielding

Safety shielding is required any time there is a risk of explosion, splash hazard or a highly exothermic reaction. All manipulations of corrosive materials which pose splash risk must occur in a chemical fume hood with the sash in the lowest feasible position. Portable shields, which provide protection to all laboratory occupants, are acceptable. Uses with potential for explosion must consult EHS for assistance in obtaining appropriate explosion shielding; chemical fume hoods do not offer sufficient protection.

This structure is acceptable when writing general rules and is very common. It does have the disadvantage mentioned in section 18.2 above: it mixes explanation with direction, a potentially-confusing situation. A structure such as that in Table 18.3 may also be used, adapting the two-column structure recommended for tasks with specific chronological steps.

Table 18.3: Excerpt from improved SOP for use of corrosive materials

DO	WHY
Safety shielding	
Use a chemical fume hood with sash lowered for work involving possible splashes of corrosives.	Violent splashes can dislodge face shields and chemical splash goggles (which must still be worn regardless of the shielding provided). The sash provides extra splash protection.
Use portable explosion shields for work involving risk of explosion. (Consult EHS for assistance.)	Chemical fume hood sashes are usually tempered glass, which is not sufficiently strong to withstand many explosions. Hood sashes can shatter and send glass shrapnel throughout the laboratory.
Safety shower	
Provide a safety drench shower/eyewash unit or eye/face/bodywash within 10 seconds uninterrupted travel of the corrosives work area.	The longer corrosives are allowed to remain on the skin or in the eyes before rinsing begins, the more likely it is that permanent tissue damage will result
Special storage	
Segregate acids from bases	Leaks (or damage) to multiple bottles could cause an energetic reaction and toxic vapor production if acids and bases mix.
Segregate liquid corrosives from solids	Liquid corrosives are more easily cleaned up by neutralizing agents if they are not mixed with solids. Additionally, spilled corrosive solids may be cleaned up more easily than liquids.
Store corrosives in plastic trays (except perchloric acid)	The plastic trays provide secondary containment to limit the spread of leaks and prevent incompatible corrosives from mixing. Trays should be deep enough to contain leaks from all bottles stored in the tray. For perchloric acid, store separately in a large glass beaker—spilled perchloric acid can react violently with some plastics.
Store corrosives **below eye level**	Storing chemicals above eye level increases the chance of dropping a bottle during retrieval. A bottle dropped from a height is more likely to break and spray corrosives around the lab (and splash the lab occupants).

Nevertheless, paragraph form is not advisable for **tasks** related to the use of hazardous materials. These operations, such as spill cleanup instructions or directions for neutralization and disposal of waste materials, should clearly separate steps to be taken from explanatory detail and peripheral issues.

18.4 Writing protocols for experimental procedures

Procedures, or more descriptively, "experimental protocols," prepared to guide safe conduct of an experiment should not only be about safety. All scientists and engineers know the value of complete planning in an experiment. Indeed, without a clear purpose and procedure, the "experiment" is nothing more than "playing in lab"—there is no hypothesis to be tested nor will the results have useful meaning.

Table 18.4 shows a protocol for conducting a Charpy material impact test, a standardized procedure for determining the **impact strength** of a material, that is, the amount of energy necessary to cause brittle failure of the material. A Charpy tester is basically a shaped weight mounted on the end of a pendulum, with the sample mounted vertically at the bottom of the pendulum's swing. To conduct the test, the pendulum is raised to a predetermined height and dropped to impact the sample. (The gravitational potential energy of the weight is thus transferred in a reproducible and easily-measured manner.) Figure 18.2 shows a Charpy tester being loaded; Figure 18.3 shows the same machine in operation.

Figure 18.2: Laboratory manager Bryan Crawford loading Charpy impact tester in neutral position.

Figure 18.3: Charpy impact tester in operation.

Table 18.4: Charpy impact test SOP

Low Point University
Department of Materials Engineering
SOP for conducting a Charpy impact test with tester in room 4
Revision 2.1
Principal author: Dr Oliver Twist

	Step		Notes
1	UNLOCK and REMOVE the chain securing the weighted arm in the neutral position	–	Place chain and lock out of range of the arm's swing so that they do not interfere.
		–	Place chain and lock away from areas of travel—do not create trip hazards.
		–	The key to the lock is available from the lab manager. You will not be given the key until you have been trained and fully qualified on this procedure.
2	PUT ON safety glasses	–	Use a CLEAN pair of safety glasses taken from the sterilizer cabinet on the lab wall.

Table 18.4: (continued)

	Step	Notes
3	RAISE the arm **slightly** and MOUNT the test sample in position against the anvil.	– Place the notch AWAY from the direction of travel of the arm (to the left as you face the apparatus). – WARNING! PINCH HAZARD: only raise the arm enough to clear the anvil. – Maintain firm control of the arm during sample placement to avoid pinching or crushing hands or fingers. – DANGER! IMPACT HAZARD: DO NOT attempt this step with the arm raised to one of the fully-latched positions or held up by another person! Your head or body may be in the path of the arm swing and you may be seriously injured or killed!
4	MOVE the selector lever atop the unit to LATCH position.	– Do not attempt to raise the arm beyond the slight amount used for mounting the sample unless the selector is in LATCH position.
5	RAISE the arm to the desired latch position	– Use the first latch position for softer materials such as aluminum and the second for harder materials such as steel. – DANGER! IMPACT HAZARD: Clear the area through which the arm will swing and maintain it clear of personnel until the test is completed and the arm comes to a complete stop; personnel may be struck and seriously injured or killed by the moving arm. – Note that this unit is NOT equipped with a guard or sensor to exclude personnel from the danger zone! – WARNING! IMPACT HAZARD: The pieces of the broken sample will be thrown from the apparatus against the nearby wall, along the line of travel of the arm. Stay out of this area to avoid being struck.
6	MOVE the dial pointers to the "0" position	– There are two co-axial pointers—one shows the instantaneous impact energy and one the maximum.
7	CALL out "CHARPY TEST READY—STAND CLEAR"	– This is not only to warn persons you can see, but also to prevent those behind you inadvertently walking into the arm's path of travel.
8	COUNT aloud down from 5 to zero	–
9	MOVE the selector lever from LATCH position to TEST position	– This will release the arm to fall and strike the sample.
10	ALLOW the arm to swing down and strike the sample	– WARNING! IMPACT HAZARD: The arm will swing back and forth several times—stay out of its line of travel to avoid being struck.

Table 18.4: (continued)

	Step	Notes
11	MOVE the selector lever from TEST position to BRAKE position and hold it there until the arm comes to rest.	– The BRAKE position is spring-loaded and will not remain latched
12	COLLECT the broken pieces of the sample from the floor between the tester and the wall	– Archive or dispose of the sample properly
13	RECORD the maximum impact energy off the dial	–
14	SECURE the weighted arm using the lock and chain	– Tighten the chain before applying the lock so that the arm can barely move.
15	TAKE OFF safety glasses	– Return the safety glasses to the box marked "USED SAFETY GLASSES;" the lab manager will sterilize them at the end of the day.

The procedure in Table 18.4 follows a classic two-column structure, giving clear directions for what is to be done and in what order. Side notes are just that: notes set to the side of the main instructions. If the instructions for carrying out the test were written in paragraph form, they would be far more difficult to understand and follow. In fact, one useful technique for ensuring that a common procedure is followed exactly is to encase the procedure in plastic sheet protectors and to use it as a checklist—marking off each step with grease pencil or the like as it is completed.

18.5 Writing protocols for use of hazardous equipment

Protocols and procedures for hazardous equipment fall in-between the two previously-mentioned applications, and hazardous equipment procedures show features of both. The manuals for most complex equipment illustrate this. Most hazardous equipment has practices and safety rules that must be followed in general, regardless of what particular activity is underway—"Never place your eye on the same level as the laser optical path unless the laser is *off* and the key removed," for example. Such instructions are often accompanied by iconography or symbols to assist in communication with those who read the manual's language poorly.

18.5 Writing protocols for use of hazardous equipment

Nevertheless, hazardous equipment also has one or more procedures that must be conducted on or with the equipment: a laser user's manual might contain procedures for aligning the laser beam, adjusting power, activating and deactivating the beam, shutting down the laser for the day, etc. Table 18.5 shows a possible alignment procedure for a laser unit as it might appear in an operating manual.

Table 18.5: Procedure for aligning laser system

Abbreviated procedure for aligning a 2-mirror laser system to place a beam on a target
1. Don laser safety goggles appropriate for the beam wavelength and power.
2. Activate the laser beam. For pulsed lasers, select continuous wave operation if possible. Always use the lowest possible setting.
3. Mount a test target card at the front end of an optical rail.
4. Roughly align the two mirrors so that the laser beam strikes the target card.
5. Manipulate the first mirror to align the beam to the desired target point.
6. Move the test target card to the other end of the optical rail, further from the mirror assembly.
7. Manipulate the second mirror to align the beam to the desired target point.
8. Move the test target card back to the front of the optical rail and verify that the beam is still aligned. Repeat the mirror manipulations until the beam is fully aligned if necessary.
9. Turn off the laser and remove laser safety goggles.

Beyond operating manuals, which are written for general instruction, hazardous equipment may have additional SOPs, written to communicate local safety protocols, such as a list of personnel allowed in the room when the equipment is operating, or to detail aspects of experimental procedure. An example of the latter would be a laser startup SOP following the one in the manual but also explaining exactly where the laser activation and emergency shutdown physically are. Table 18.6 is a lightly-edited example of an actual SOP used in a laser installation.

Table 18.6: Sample laser SOP

Laser Standard Operating Procedure
SOP—HJB001

Principal Investigator	Dr. Hugh James Bissel	Date	2016-03-30
Laser Safety Officer	Charles H. Townes	Date	2016-03-31

Experimental purpose:
This procedure is for the ablation and machining of metallic samples. The samples are irradiated with 532 nm coherent light and material is removed by ablation.

Table 18.6: (continued)

All persons who use lasers listed herein must read and sign this SOP annually.

Laser Safety Contacts
Laser Safety Officer: Charles H. Townes
Office: Room 532 Coherence Building
Phone 410-555-1234
Principal Investigator: Dr. Hugh James Bissel
Office: Room 721 Nanometer Hall
Phone: 410-555-5678

Medical Emergency: Call 911

Emergency Procedure;
1. Confirm laser is shut down or in safe mode
2. Call 911
3. Provide operator with
   a. Nature of injury (if laser eye injury say so explicitly)
   b. Current location
   c. Name and general condition of persons injured

General Laser Safety Program
Operator Responsibilities

Operators of the laser systems covered by this SOP shall:
– Attend appropriate training before working with lasers;
– Use lasers safely;
– Ensure they are in compliance with established policy and procedures;
– Promptly report any malfunction, problem, incident, or injury that may impact safety.

Posting Requirements
The laser workspace shall be posted with appropriate signs indicating the nature of the hazard. Signs shall be specified by the Laser Safety Officer in conformance with ANSI Z136.1 guidelines.

Laser Systems In Use

Class	4
Type	Diode
Manufacturer	Coherent
Model No.	Talisker 1000 532-15
Serial No.	10002015051110
Wavelengths (nm)	532
Beam Diam. (mm)	2.2
Power (W)	15
CW or Pulse	Pulse
Pulse Width	<15 ps
Location	Room 140 Nanometer Hall

Table 18.6: (continued)

Authorized Personnel

Dr. Hugh James Bissel (PI), Mark Welby (operator/graduate student), Oliver Twist (operator/postdoctoral fellow)

Operating Procedures

Initial Laboratory Preparation

1. Ensure that the laboratory door is closed.
2. Ensure laser warning system is active.

Identification of Personnel

Ensure that all personnel present are:

- On the Authorized Personnel list (or are escorted by an Authorized Person);
- Have completed appropriate laser training;
- Are aware that the laser is about to be activated;
- Are following the SOP and administrative controls specified for the laser.

Don Personal Protective Equipment

1. Inspect laser safety eyewear.
2. Don laser safety eyewear.

Laser Turn-On Procedure

1. Insert sample into sample slot.
2. Check beam enclosure is closed and latched.
3. Check beam shutter is closed.
4. Insert key into key lock. Turn the key lock and ensure the red LEDs turn on. Abort the operation and troubleshoot if the red LEDs do not light.
5. Check that the computer control panel indicates beam on. Abort the operation and troubleshoot if there is no "BEAM ON" indication.
6. Select "START" on computer control panel to begin the machining program.
7. Confirm that laser warning light switches from Safe to Active.

Laser Turn-Off Procedure

1. Machining will stop automatically when program ends.
2. Turn key and ensure red LEDs are off. If red LEDs do not turn off, shut down and lock out power at the electrical service disconnect for the laser before troubleshooting.
3. Open the enclosure and remove the sample.
4. Turn off the laser warning light.
5. Remove laser safety eyewear.
6. Confirm automatic laser warning signs have reverted to read 'Safe.'

Table 18.6: (continued)

Non-beam Hazards

- 120 V/10 A electrical power supply.
- Laser-generated air contaminants: depend on materials being processed; review materials against allowed inventory list before processing.

Controls
Engineering Beam Controls
Beam operated only inside engineered and interlocked beam enclosure. As the beam enclosure interlock is defeatable (e.g., by removing the laser from its mount), laser safety eyewear is required although technically the system is Class 1. This experimental device is not cleared for distribution with regulatory authorities and may not be removed from campus.

Administrative Controls
The laser shall not be operated outside the custom enclosure.

Personal Protective Equipment
YRD #35 LaserBlockers, Green Lenses, supplied by Laser Blockers International.

19 Annotated bibliography of laboratory safety references

The list below comprises a small set of reference books and papers that the author finds useful on a regular basis. Some are elementary texts, such as the "Safety in academic chemistry laboratories" booklet by the American Chemical Society, while others are primary sources such as the American National Standards series on laser safety.

- **20% Chance of rain** [122]. A discussion of the concept of risk, its estimation, and its perception, couched in language understandable to any researcher regardless of field.
- **ACS Handbook of chemical health and safety** [123]. The **ACS Handbook** contains a series of articles covering a broad range of chemical hazards, from peroxidizable organic compounds to waste disposal and management. Highly recommended for the chemical professional.
- **ANSI Z136.1 (Safe Use of Lasers)** [43]. This is the primary American standard for laser safety, and it contains both requirements and recommendations for maintaining a safe laser facility. It also contains detailed calculation methods for determining the extent of the Nominal Hazard Zone for the installation, as well as for selection of protective eyewear. Several sub-standards (Z136.2, .3, etc.) exist for specific environments; most notable among these is Z136.7, specialized for research environments.
- **Biosafety in microbiological and biochemical laboratories** [90]. The "BMBL" is the "Bible of biosafety." It comprises two main parts: the first is a complete discussion of biological risk factors, biosafety containment levels, and appropriate safe biological practices; while the second part is a fairly comprehensive listing of biological agents along with recommendations for appropriate practices and containment levels.
- **Bretherick's handbook of reactive chemical hazards** [78]. Bretherick's handbook is the foremost reference book on chemical reactivity and compatibility. It is essentially an annotated guide to literature reports of chemical reactivity issues: generally, it gives a short (or for some compounds, not so short) summary of hazards and interactions, heavily footnoted, with complete literature references following.
- **Electrical safety: Systems, sustainability, and stewardship** [124]. Comprehensive discussion of electrical safety issues, from generating equipment through grounding. Special chapters cover items such as lockout/tagout regulations.
- **Emergency care for hazardous materials exposure** [66]. This book is not so much intended for the researcher but for the health professional. Emergency medical technicians working for laboratory facilities may find the treatment protocols useful in their work. Laboratories handling exotic (highly toxic,

pyrophoric, etc.) compounds would be well-served to ensure that this text was available to the staff of the nearest emergency department.
- **Identifying and evaluating hazards in the research laboratory** [112]. An extensive discussion of hazard analysis methods appropriate to laboratory operations. Particular attention is given to specialized methods such as "control banding" suitable for chemical operations.
- **Inherently safer design: An overview of key elements** [1]. A discussion of the principles of Inherently Safer Design in the specific context of chemical process design.
- **Job Safety Analysis: A practical tool for ensuring safety of the workplace** [125]. A paper from the African Newsletter on Occupational Safety and Health that clearly and concisely explains the JSA technique.
- **Laser safety guide** [109]. A small booklet containing the rudiments of laser safety, aimed at the first-time user. The author has found this booklet useful as a text for initial laser awareness training.
- **Learning by Accident** [126]. A 3-volume set of spiral-bound laboratory accident descriptions covering both K-12 and college/university level situations.
- **Managing the risks of organizational accidents** [23]. A classic discussion of the organizational issues that lead to accidents, focusing on root causes rather than the immediate factors that led to the accident. Since appropriate response to accidents includes both fixing the individual issues involved in the incident as well as addressing problems in policies, procedures, and culture that created those issues.
- **Prudent practices in the laboratory: Handling and management of chemical hazards** [120]. **Prudent Practices**, as it is called, is one of the two chemical hazard references that the author would recommend for any laboratory (the other being the ACS Handbook of chemical health and safety, above). In addition to information on the chemical hazards themselves, Prudent Practices provides a wealth of information on safe laboratory operation, as well as sample SOPs and an extensive inspection checklist. It also provides some information on physical hazards often encountered in the chemical laboratory (e.g., cryogens).
- **Safety in academic chemistry laboratories: Accident prevention for college and university students** [60]. This booklet, part one of a set of two (the second is intended for administrators and faculty) is a very succinct complication of the most important facts and practices for use in chemistry laboratories, both teaching and research. Appropriate to its origin in the American Chemical Society (ACS), the booklet concentrates almost exclusively on laboratory practices relevant to chemical use—chemical hazards of the materials, appropriate PPE, etc., with some peripheral discussion of chemistry-related lab physical hazards such as sharps, steam baths, etc. An electronic version of the book is currently available for free from the ACS website (http://www.acs.org).

- **The checklist manifesto: How to get things right** [111]. A short business book arguing for the increased use of checklists in ensuring procedures are done consistently and correctly. The book contains substantial advice on writing effective checklists.
- **The dose makes the poison** [127]. A thorough discussion of toxicology for the non-specialist and educated layperson. Those who work with toxic chemicals should read this book to deepen their understanding of the meaning of "toxic" and the variety of toxic effects chemicals may have.
- **What Went Wrong? Case histories of process plant disasters and how they could have been avoided** [128]. This book addresses industrial rather than laboratory practice, giving hundreds of accident descriptions along with analyses of varying degrees. Most of the types of failures that cause industrial accidents exist in the laboratory environment, so it is a valuable reference for the student of laboratory safety.
- **WHO Laboratory biosafety manual** [92]. The World Health Organization's equivalent of the "BMBL," above. Useful for its explanation of BSL—1–4 construction practices and clear explanations of how various types of biological safety cabinet function.

References

[1] Hendershot D. Inherently safer design: An overview of key elements. Prof Safety 2011, 56, 48.
[2] Marquardt A. Exploding chewing gum kills student. ABC News, 2009. (Accessed 2015-06-25 at http://abcnews.go.com/International/chewing-gum-explodes-killing-student-ukraine/story?id=9290557.)
[3] Nierenberg DW, Nordgren RE, Chang MB et al. Delayed cerebellar disease and death after accidental exposure to dimethylmercury. N Engl J Med 1998, 338, 1672–1676.
[4] Alcedo JA, Wetterhahn KE. Chromium toxicity and carcinogenesis. Int Rev Exp Pathol 1990, 31, 85–108.
[5] Witt SF. Dimethylmercury. US Occupational Safety and Health Administration, 1991.
[6] Endicott K. The trembling edge of science. Dartmouth Alumni Magazine 1998-04.
[7] Blayney MB, Winn JS, Nierenberg DW. Handling dimethylmercury. Chem Eng News 1997, 75, 7.
[8] Edwards GN, Cantas MD. Two cases of poisoning by Mercuric Methide. St Bartholomew's Hospital Reports 1865, 1, 141–149.
[9] Foderaro LW. Yale student killed as hair gets caught in lathe. New York Times 2011; A, 19.
[10] Move D. Monica Thayer loses scalp in Ohio industrial accident. Huffington Post, 2012. (Accessed 2016-03-02 at http://www.huffingtonpost.com/2012/07/26/monica-thayer-loses-scalp-ohio-accident_n_1706432.html.)
[11] McLaughlin TP, Monahan SP, Pruvost NL, Frolov VV, Ryanazov BG, Sviridov VI. A review of criticality accidents. Los Alamos, NM, US Los Alamos National Laboratory, 2000.
[12] Schreiber R. Memorandum. 1946;
[13] Malenfant RE. The Accident. American Nuclear Society 1995 Annual Meeting. 1995.
[14] Malenfant RE. Personal communication. 2015.
[15] National Council on Radiation Protection and Measurements. Limitation of exposure to ionizing radiation. Bethesda, MD, US, National Council on Radiation Protection and Measurements, 1993.
[16] Lapp R. Atoms and people. New York, NY, US, Harper and Bros., 1957.
[17] US Chemical Safety and Hazard Investigation Board. Texas Tech University: Laboratory explosion. Washington, DC, US, US Chemical Safety and Hazard Investigation Board, 2010.
[18] Johnson J, Kemsley J. Academic lab safety under exam. Chem Eng News 2011, 89, 25–27.
[19] Aldrich Chemical Company. Technical bulletin AL-164: Handling pyrophoric reagents. Milwaukee, WI, US, Aldrich Chemical Company, 1995-06.
[20] Baudendistel B. Investigation report, Case number S1110—003—09. Los Angeles, CA, US, Department of Industrial Relations, Division of Occupational Safety and Health, Bureau of Investigations, State of California, 2009.
[21] Kemsley J. Learning from UCLA. American Chemical Society, 2009. (Accessed 2015-07-03 at http://cen.acs.org/articles/87/i31/Learning-UCLA.html.)
[22] Lacey JW, Williams, SP; Hum, CW, Rizzo, M. People of the State of California v. Patrick Harran: Deferred prosecution agreement. Los Angeles, CA, US, Superior Court of the State of California for the County of Los Angeles, 2014.

[23] Reason JT. Managing the risks of organizational accidents. Burlington, VT, US, Ashgate Publishing Company, 1997.
[24] Rebbett D. Pyramid power: A new view of the Great Safety Pyramid. Prof Safety 2014, 59, 30–34.
[25] Peart N. Dreamline. Roll the Bones. Los Angeles, CA, US, Atlantic Records, 1994.
[26] American Institute of Chemical Engineers. Code of Ethics. New York, NY, US, American Institute of Chemical Engineers. (Accessed 2015-07-30 at http://www.aiche.org/about/code-ethics.)
[27] American Society of Mechanical Engineers. Society policy: Ethics. New York, NY, US, American Society of Mechanical Engineers, 2012. (Accessed 2015-07-30 at https://www.asme.org/getmedia/9EB36017-FA98-477E-8A73-77B04B36D410/P157_Ethics.aspx.)
[28] Royal Society of Chemistry (UK). Code of conduct. Cambridge, UK, Royal Society of Chemistry (UK), 2012. (Accessed 2015-07-30 at http://www.rsc.org/images/code-of-conduct_tcm18-5101.pdf.)
[29] Corts TE. Bliss and Tragedy: The Ashtabula Railway-Bridge Accident of 1876 and the Loss of P.P. Bliss. Birmingham, AL, US, Samford University Press, 2003. Ref. Appendix C, "Transcription of the handwritten report of Dr Frank H. Hamilton, June 1878, upon his examination of the remains of Charles Collins."
[30] Kuespert DR. Delegating safety. New York, NY, US, American Institute of Chemical Engineers Center for Chemical Process Safety Sixteenth Annual Conference, 2001.
[31] Harris CE, Pritchard MS, Rabins MJ, James R, Englehardt E. Engineering ethics: Concepts and cases. Boston MA, US, Wadsworth Cengage Learning, 2014.
[32] Agence-France Presse. Explosion à l'école de chimie de Mulhouse: un professeur tué. Agence-France Presse, 2006. (Accessed 2016-05-23 at http://lci.tf1.fr/france/2006-03/explosion-ecole-chimie-mulhouse-professeur-tue-4856984.html.)
[33] Laurent A, Perrin L, Dufaud O. Investigation and analysis of an explosion in a research laboratory at a French university. Chemical Engineering Transactions 2014, 36, 7–12.
[34] Agence-France Presse. Explosion: un professeur jugé à Mulhouse. Agence-France Presse, 2010. (Accessed 2016-03-03 at http://www.lefigaro.fr/flash-actu/2010/09/13/97001-20100913FILWWW00590-explosion-un-professeur-juge-a-mulhouse.php.)
[35] Cheval A. Explosion mortelle à l'Ecole de Chemie: Peine Confirmée de 18 mois avec sursis en appel. 2011. (Accessed 2016-05-23 at http://www.thiesvision.com/Mulhouse-Explosion-Mortelle-a-L-Ecole-de-Chemie-Peine-Confirmee-de-18-mois-Avec-Sursis-En-Appel_a1113.html.)
[36] Lobjooie G. School of chemistry—Professor sentenced to 18 months suspended. Paris, FR, FENVAC (Fédération national des victims d'attentats et d'accidents collectifs; French national federation for the victims of disasters), 2010.
[37] Kuespert DR, Leon NJ. A generalized risk assessment system for laboratory experiments. Paper in preparation.
[38] Center for Chemical Process Safety. Guidelines for hazard evaluation procedures. New York, NY, US, American Institute of Chemical Engineers, 2008.
[39] Buzan T. The mind map book: How to use radiant thinking to maximize your brain's untapped potential. New York, NY, US, Penguin Books, 1993.
[40] von Oech R. A whack on the side of the head: How you can be more creative. New York, NY, US, Warner Books, 2008.

[41] Pronovost PJ, Vohr E. Safe patients, smart hospitals: how one doctor's checklist can help us change health care from the inside out. New York, NY, US, Hudson Street Press, 2010.
[42] American Society of Heating, Refrigerating, and Air-conditioning Engineers. ANSI/ASHRAE 15–2013, Safety standard for refrigeration systems. Atlanta, GA, US, American Society of Heating, Refrigerating, and Air-conditioning Engineers, 2013.
[43] Laser Institute of America. ANSI Z136.1–Safe use of lasers. Orlando, FL, US, Laser Institute of America, 2014.
[44] US Occupational Safety and Health Administration. Machine guarding eTool. US Occupational Safety and Health Administration. (Accessed 2016-03-03 at https://www.osha.gov/SLTC/etools/machineguarding/.)
[45] National Fire Protection Association. ANSI/NFPA 51B: Standard for fire prevention during welding, cutting, and other hot work. Quincy, MA, US, National Fire Protection Association, 2014.
[46] Air Products and Chemicals. Safetygram 22: Liquid helium. Air Products and Chemicals, 2014.
[47] Gupta S, Jani CB. Oxygen cylinders: "life" or "death?" Kampala, UG, Afr Health Sci 2009, 9, 57–60.
[48] Canadian Centre for Occupational Health and Safety. Cleaning with compressed air. Canadian Centre for Occupational Health & Safety. (Accessed 2015-07-15 at http://www.ccohs.ca/oshanswers/safety_haz/compressed_air.html.)
[49] Cook L. How to spot dangerous air and fluid injection injuries. J Emerg Med Services, 2009-04-13. (Accessed 2015-07-15 at http://www.jems.com/articles/2009/04/how-spot-dangerous-air-and-flu.html.)
[50] Kale TR, Momin M. Needle-free injection technology: An overview. Innovations in Pharmacy 2014, 5, 1–8.
[51] Fisher HG, Forrest HS, Grossel SS et al. Emergency relief design using DIERS Technology: The Design Institute for Emergency Relief Systems (DIERS) Project Manual. New York, NY, US, The Design Institute for Emergency Relief Systems of the American Institute of Chemical Engineers, 1993.
[52] Hazards of the Electrical Profession. Electrical review and western electrician 1909, 54, 122.
[53] Kouwenhoven WB. Human safety and electrical shock. Research Triangle Park, NC, US, International Society of Automation (formerly Instrument Society of America), 1968.
[54] US Occupational Safety and Health Administration. Controlling electrical hazards. US Occupational Safety and Health Administration. (Accessed 2015-09-25 at https://www.osha.gov/Publications/3075.html.)
[55] Gorey M. Mal death ruled an accident. The Cleveland Stater 2005-10-14.
[56] US National Institute for Occupational Safety and Health. NIOSH Alert: Preventing worker deaths from uncontrolled release of hazardous electrical, mechanical, and other types of injury. Cincinnati, OH, US, US National Institute for Occupational Safety and Health, 1999.
[57] World Health Organization. Electromagnetic fields. Geneva, CH, World Health Organization. (Accessed 2015-07-24 at http://www.who.int/peh-emf/en/.)
[58] Nave CR. BCS Theory of Superconductivity. Atlanta, GA, US, Georgia State University Department of Physics and Astronomy, 2016. (Accessed 2016-05-21 at http://hyperphysics.phy-astr.gsu.edu/hbase/solids/bcs.html.)

[59] Wallace DC, McQueen TM. New honeycomb iridium(v) oxides: $NaIrO_3$ and $Sr_3CaIr_2O_9$. Dalton Trans 2015, 44, 20344–20351.
[60] American Chemical Society. Safety in academic chemistry laboratories: Accident prevention for college and university students. Washington, DC, US, American Chemical Society, 2003.
[61] Hatayama HK, Chen JJ, de Vera ER, Stephens RD, Storm DL. A method for determining the compatibility of hazardous wastes. Cincinnati, OH, US, US Environmental Protection Agency, 1980.
[62] Stanford University. Stanford University compatible storage group classification system. Stanford University, 2009. (Accessed 2016-03-04 at https://web.stanford.edu/dept/EHS/prod/researchlab/lab/chemstorage.pdf.)
[63] US Fire Administration. Residential structure and building fires. Washington, DC, US, US Federal Emergency Management Administration, 2008.
[64] Stone A. Personal communication. 2016.
[65] National Fire Protection Association. ANSI/NFPA 2112-2012: Standard on flame-resistant garments for protection of industrial personnel against flash fire. Quincy, MA, US, National Fire Protection Association, 2012.
[66] Currance PL, Clements B, Bronstein AC. Emergency care for hazardous materials exposure. St. Louis, MO, US, Mosby Jems/Elsevier, 2007.
[67] Kimberly-Clark Worldwide. Kimberly-Clark nitrile glove chemical resistance guide. Kimberly-Clark Worldwide, Inc., (Accessed 2016-03-04 at http://www.na.kccustomerportal.com/Documents/Upload/Application/2811/Learning%20Center/Article/K4556_10_01%20Ntrl_Chem_pstr_v3.pdf.)
[68] Gloves B. Chemrest. Best/Showa Gloves, (Accessed 2016-03-04 at http://www.chemrest.com.)
[69] Clinton M. API toxicological review: Benzene. Washington, DC, US, American Petroleum Institute, 1948.
[70] Calabrese EJ, Blain RB. The single exposure carcinogen database: Assessing the circumstances under which a single exposure can cause cancer. Toxicological Sciences 1999, 50, 169–185.
[71] Bren L. Frances Oldham Kelsey: FDA reviewer leaves her mark on history. FDA Consumer 2001.
[72] Popendorf W. Vapor pressure and solvent hazards. American Industrial Hygiene Association J, 45, 719–726.
[73] Schaefer JA. An algorithm for determining the potential for chemical exposure in a laboratory setting. Chemical Health & Safety 2000, 7, 11–13.
[74] Wilson DW, Baker ML, Hart BR, Marra JE, Schwantes JM, Shoemaker PE. Waste Isolation Pilot Plant Technical Assessment Team Report SRNL—RP—2014—01198. Aiken, SC, US, Savannah River National Laboratory,
[75] Autoacceleration. In: Gooch JW, ed. Encyclopedic Dictionary of Polymers. New York, NY, US, Springer, 2010:55–56.
[76] United States Senate Judiciary Committee. Texas City Disaster: Report to accompany S. 1077. 1955; 20.
[77] Routley JG. Fire and explosions at rocket fuel plant: Henderson, NV (May 4, 1988). US Fire Administration, Federal Emergency Management Agency, 1988.
[78] Urben PG, Battle LA, Bretherick L, Pitt MJ. Bretherick's handbook of reactive chemical hazards: an indexed guide to published data. Oxford, UK, Butterworth-Heinemann, 1995.

[79] National Fire Protection Association. NFPA 704: Standard System for the identification of the hazards of materials for emergency response. Quincy, MA, US, National Fire Protection Association, 2012.

[80] United Nations. Globally harmonized system of classification and labelling of chemicals (GHS). New York, NY, US & Geneva, CH, United Nations, 2013.

[81] Surampudi S, Yeh M-L, Siegler MA et al. Increased carrier mobility in end-functionalized organosilanes. Chem Sci 2015, 6, 1905–1909.

[82] Long G, Meek ME, Lewis M. Concise international chemical assessment document 31: N,N-dimethylformamide. Geneva, CH, World Health Organization, 2001.

[83] Goode BL, Eskin JA, Wendland B. Actin and endocytosis in budding yeast. Genetics 2015, 199, 315–358.

[84] Alberts B, Johnson A, Lewis J et al. Molecular biology of the cell. New York, NY, US, Garland Science, 2014.

[85] Tsien RY. The green fluorescent protein. Ann Rev Biochem 1998, 67, 509–544.

[86] McCann J, Ames BN. Detection of carcinogens as mutagens in the Salmonella/microsome test: assay of 300 chemicals. Proceedings of the National Academy of Science (PNAS) 1975, 72, 5135–5139.

[87] Nippon Genetics Europe. Midori Green Advance DNA Stain—Safety report. (Accessed 2016-03-03 at http://www.nippongenetics.de/download/Dokumente/03-DNA-RNA-Electrophoresis/Midori/Advance/Safety Report/Safety Report - MGA.pdf.)

[88] Sigma-Aldrich Chemical. Ethidium Bromide SDS version 4.6. 2014. (Accessed 2016-03-03 at http://www.sigmaaldrich.com/catalog/product/sigma/e7637?lang=en&region=US.)

[89] Centers for Disease Control and Prevention. Report on the inadvertent cross-contamination and shipment of a laboratory specimen with influenza virus H5N1. Atlanta, GA, US, US Centers for Disease Control and Prevention, 2014.

[90] US Centers for Disease Control and Prevention. Biosafety in Microbiological and Biomedical Laboratories. Atlanta, GA, US, US Department of Health & Human Services, Public Health Service, Centers for Disease Control and Prevention, National Institutes of Health, 2009.

[91] US Centers for Disease Control and Prevention. Herpes B Virus. US Centers for Disease Control and Prevention, (Accessed 2016-03-05 at http://www.cdc.gov/herpesbvirus.)

[92] World Health Organization. Laboratory Biosafety Manual. Geneva, CH, World Health Organization, 2004.

[93] Lennette EH. Panel V-common sense in the laboratory: recommendations and priorities. Biohazards in Biological Research 1973, 353.

[94] Barker K. At the bench: A laboratory navigator. Cold Spring Harbor, NY, US, Cold Spring Harbor Press.

[95] Barbano MF, Wang H-L, Morales M et al. Feeding and reward are differentially induced by activating GABAergic lateral hypothalamic projections to VTA. J Neurosci 2016, 36, 2975–2985.

[96] Saunders BT, Richard JM, Janak PH. Contemporary approaches to neural circuit manipulation and mapping: focus on reward and addiction. Phil Trans R Soc B 2014, 370

[97] Keiflin R, Janak PH. Dopamine prediction errors in reward learning and addiction: From theory to neural circuitry. Neuron Rev 2015, 88, 247–263.

[98] Bohutskyi P, Liu K, Kahled Nasr L et al. Bioprospecting of microalgae for integrated biomass production and phytoremediation of unsterilized wastewater and aerobic digester centrate. Appl Microbiol 2015, 99, 6139–6154.

[99] Mukherjee B. Radiation shielding of a 230 MeV proton cyclotron for cancer therapy. 2009. (Accessed 2015-09-01 at http://www-mpy.desy.de/AccPhySemDESY/y2009/BhMukherjee_RadShielding230GeVp.pdf.)

[100] Peller S. Lung cancer among mine workers in Joachimsthal. Human Biology 1939, 11, 130–143.

[101] US Nuclear Regulatory Commission. Radiation exposure and cancer. US Nuclear Regulatory Commission, 2014. (Accessed 2015-09-04 at http://www.nrc.gov/about-nrc/radiation/health-effects/rad-exposure-cancer.html.)

[102] International Organization for Standardization (ISO). ISO 21348—Space environment (natural and artificial)—Process for determining solar irradiances. International Organization for Standardization (ISO), 2007.

[103] National Institute for Occupational Safety and Health. Criteria for a recommended standard: Occupational exposure to ultraviolet radiation. Atlanta, GA, US, National Institute of Occupational Safety and Health, 1973.

[104] Baan R, Grosse Y, Lauby-Sécretan B et al. Carcinogenicity of radiofrequency electromagnetic fields. The Lancet Oncology 2011, 12, 624–626.

[105] Federal Communications Commission Office of Engineering and Technology. Questions and answers about biological effects of radiofrequency electromagnetic fields. Washington, DC, US, Federal Communications Commission, 1999.

[106] Federal Communications Commission Office of Engineering and Technology. Evaluating compliance with FCC guidelines for human exposure to radiofrequency electromagnetic fields. Washington, DC, US, US Federal Communications Commission, 1997.

[107] Chaffee JG. Method and apparatus for heating dielectric materials. US Patent 2,147,689. 1939.

[108] Hadler J, Tobares EL, Dowell M. Random testing reveals excessive power in commercial laser pointers. J Laser Applications, 2013, 25.

[109] Barat K. Laser safety in the lab. Bellingham, WA, US, Spie Press, 2010.

[110] Fischer M, Fuerst B, Lee SC et al. Preclinical usability study of multiple augmented reality concepts for K-wire placement. Int J Computer Assisted Radiology and Surgery, 2016, 1–8.

[111] Gawande A. The checklist manifesto: How to get things right. New York, NY, US, Picador, 2009.

[112] ACS Committee on Chemical Safety. Identifying and evaluating hazards in research laboratories. Washington, DC, US, American Chemical Society, 2015.

[113] Card AJ, Ward AR, Clarkson PJ. Beyond FMEA: The structured what-if technique (SWIFT). J Healthcare Risk Manag 2012, 31, 23–29.

[114] Delstar Metal Finishing. Electropolishing: A User's Guide to Applications, Quality Standards, and Specificiations. Houston, TX, US, Delstar Metal Finishing, 2003.

[115] Chen W, Song B. Split Hopkinson (Kolski) bars: Design, testing, and applications. New York, NY, US, Springer Science & Business Media, 2011.

[116] UK Health and Safety Executive. Flixborough (Nypro UK) explosion 1st June 1974. Health and Safety Executive, UK, (Accessed 2016-01-014 at http://www.hse.gov.uk/comah/sragtech/caseflixboroug74.htm.)

[117] Kletz T. What you don't have, can't leak. Chemistry and industry 1978, 6, 287–292.
[118] Shrivastava P. Bhopal: Anatomy of a crisis. Cambridge, MA, US, Ballinger Pub. Co., 1987.
[119] US Chemical Safety and Hazard Investigation Board. Key lessons for preventing incidents from flammable chemicals in educational demonstrations. Washington, DC, US, US Chemical Safety and Hazard Investigation Board, 2014.
[120] National Research Council of the National Academies Committee on Prudent Practices in the Laboratory: An Update. Prudent practices in the Laboratory: Handling and management of chemical hazards. Washington, DC, US, The National Academies Press, 2011.
[121] Compressed Gas Association. CGA handbook of compressed gases. Chantilly, VA, US, Compressed Gas Association, 2013.
[122] Jones RB. 20% chance of rain: Exploring the concept of risk. Hoboken, NJ, US, Wiley, 2012.
[123] Alaimo RJ. Handbook of chemical health and safety. Washington, DC, US, American Chemical Society, 2001.
[124] Boss MJ, Nicoll G. Electrical safety: Systems, sustainability, and stewardship. Boca Raton, FL, US, CRC Press, 2015.
[125] Kiwakete HM. Job Safety Analysis: A practical tool for ensuring safety of the workplace. Afr Newsletter on Occ Health and Safety 2008, 18, 36–37.
[126] Laboratory Safety Institute. Learning by accident. Natick, MA, US, Laboratory Safety Institute, 1997.
[127] Frank P, Ottoboni MA. The dose makes the poison. Hoboken, NJ, US, Wiley, 2011.
[128] Kletz T. What went wrong: Case histories of process plant disasters and how they could have been avoided. Burlington, MA, US, Elsevier, 2009.

Index

A

absorption 86, 89, 91, 107, 108, 153, 155, 156
accident prevention 248
accidents 3, 9, 10, 14, 15, 20, 22, 26, 31, 74, 99, 104, 111, 175, 186, 208, 230, 248, 249
ACGIH (American Council of Governmental Industrial Hygienists) 107, 108
acids 91, 96, 97, 100, 104, 199, 215, 216, 232, 233
acids and bases 91, 100, 104, 232, 233
ACS. *See* American Chemical Society 247, 248
administrative controls 6, 24, 30, 31, 43, 160, 161, 165, 186, 194, 245
aerosolization 127, 145
aerosols 129, 132, 135, 136, 141, 143
air 6, 9, 15, 19, 38, 40, 47, 52, 53, 56, 59, 63, 64, 66, 67, 71, 75, 76, 84, 87, 90, 93, 95, 96, 98, 102, 111, 115, 120, 130–132, 136–142, 149, 159, 193, 209, 224, 227, 232, 246
airflow 89, 131, 138, 139, 141, 142
air guns 63
American Chemical Society (ACS) 248
American Council of Governmental Industrial Hygienists (ACGIH) 107
American National Standards Institute (ANSI)
ampoules 52, 57, 94, 225
animals 91, 106, 119, 129, 131, 155, 156, 215
arc 74, 75, 76, 99
asphyxiation 57, 59, 67, 84
autoclave 62, 132–134, 136, 141
autoclaving 132, 134, 136, 179

B

beta particles 149
biohazard 51, 129, 142, 143, 147, 179, 231
biohazardous 51, 125, 145, 172, 173, 179, 227
biological materials 125, 129, 131, 134, 136, 137
biological safety cabinet. *See* BSC 63, 73, 94, 130–132, 135–141, 146, 154, 217, 221, 222, 228, 229, 231, 249
biology 49, 105, 121, 123, 143, 144, 147
Biosafety Level. *See* BSL 126, 127, 129–133, 147
biosafety professionals 126, 127, 133
blast shield 8, 83, 84, 87
blood 63, 67, 79, 81, 118, 126, 144, 145, 155, 158

bloodstream 49, 63, 86, 100, 144
boiloff vapor 59, 63, 71, 72, 216
brainstorming 28, 31, 35, 39, 180, 182
breakdown, dielectric 75, 76, 105, 123, 158, 224
Bremsstrahlung 148, 150, 151
broken glass 20, 50, 52, 86
BSC (biological safety cabinet) 136–143, 146, 228, 229, 231, 232
BSL (Biosafety Level) 126–134, 145, 146, 249
Bunsen burner 52
burn 25, 44, 54, 91, 93, 99, 113
burners 52, 54, 94, 142, 143, 210, 216

C

cabinet 25, 63, 80, 87–89, 91, 95, 104, 109, 130, 136–143, 146, 173, 174, 216, 227, 231, 240, 249
– corrosives 71, 90, 91, 104, 172, 173, 216, 232, 233
– flammables 54, 90, 94–97, 99, 104, 139, 172, 173, 216, 231
cancer 15, 48, 53, 90, 105, 106, 109, 149, 152, 175, 179, 215
carcinogens 90, 105, 108
cause 6, 10–12, 15, 19, 50, 51, 53, 54, 59, 71, 74–76, 79, 81, 89, 90, 94, 96, 98, 99, 100, 105, 106, 112, 117, 119, 125, 126, 128, 129, 132, 135, 139, 143, 145, 152, 154, 157–159, 166, 175, 177, 178, 185, 186, 192, 207, 232, 233, 239, 249
– root 11, 86, 117, 248
cells 48, 100, 105, 119, 148, 179
Charpy tester 239
checklist 11, 12, 23, 26, 28, 36, 39, 171–173, 175, 180, 181, 215, 216, 223–225, 227, 232, 248, 249
checklist-form hazard review 224
checklist technique 171, 223
checklist zombies 12
chemical exposures 107
chemical fume hood 4, 25, 54, 63, 66, 85, 87–89, 94, 96, 108, 124, 137, 139, 141, 199, 216, 232, 233
chemical handling 90
chemical hazard 86, 87, 115, 118, 122, 248

chemical hazards, reactive 13, 26, 47, 86, 91, 102, 110–113, 118, 120, 159, 182, 215, 216, 247, 248
chemical incident 113
– reactive 26, 91, 102, 110–113, 247
chemical reactions 53, 55, 65, 72, 94, 95, 110–112
chemicals 3, 12–17, 44, 47, 57, 63, 86, 87, 89, 90, 91, 94–97, 99, 101, 102, 104–111, 113, 114, 116, 118, 124, 125, 128, 139, 151, 162, 172, 173, 175, 177, 199, 200, 208, 215, 216, 225, 232, 233, 249
– flammable 9, 19, 37, 54, 56, 67, 71, 91, 93–97, 99, 104, 113–115, 117, 118, 120, 124, 139, 143, 172–174, 177, 193, 208, 215, 216, 224, 225, 227, 231
– highly-hazardous 108, 109, 210
– oxidizing 54, 60, 67, 96, 97, 100, 104, 111–113, 199, 216, 224
chemistry 3, 4, 7, 9, 19, 44, 48, 49, 82, 83, 90, 97, 107, 118, 121, 176, 201, 203, 211, 247, 248
chemistry laboratories 90, 121, 211, 248
chips, flying 11, 27–29, 47, 49, 63, 73, 174
chronic effects 105
chronic health hazard 115
clean air workstation 140
clobbering 47, 220
clothing 44, 54, 61, 76, 99, 154, 161, 211
codes 15, 71, 97, 110, 175, 232
combustible materials 52, 54, 96, 97, 154, 225
combustibles 54, 76, 84, 85, 92
combustion 67, 71, 93, 96, 192
community risk 131, 132
compounds, active metal 4, 8, 13, 25, 53, 82, 85, 86, 87, 97–100, 102, 104, 105, 107, 111, 112, 114, 118–121, 139, 172, 173, 199, 247, 248
compressed air 56, 63, 67, 71, 208
Compressed Gas Association 71, 172
compressed gas cylinders 54, 63, 66, 67, 81, 97
compressed gases 14, 63, 66, 67, 71, 172, 173, 224
consequences 17, 18, 20, 23, 28–31, 38, 181, 185–187, 189, 191–194, 203
containment 25, 52, 64, 91, 98, 104, 110, 129–132, 134, 146, 202, 203, 216, 233, 247
contaminants 89, 90, 112, 137, 138, 141, 246
contamination 57, 63, 73, 97, 103, 109, 128, 134–136, 140–142, 227

cooling 37, 54, 55, 56, 58, 59, 63, 72, 182, 199
cord, extension 35, 76, 77, 81, 84, 85, 216
cornea 153, 157, 158
corrosive 95, 100, 102–104, 110, 114, 143, 172, 173, 216, 232, 233
cosmic rays 148, 150
creativity in hazard control 31
credible case, worst 27, 177, 199
cryogenic freezers 136
cryogens 58, 59, 61–63, 67, 71, 215, 216, 248
cultures 44, 82, 130, 134, 136, 139, 143, 179, 227

D

damage, hearing 3, 13, 18–20, 24, 26, 29, 35, 49, 63, 71, 74, 76, 77, 79, 81, 95, 100, 104, 105, 118, 119, 125, 126, 135, 145, 154, 156–158, 203, 224, 233
death 3–5, 9, 15, 20, 22, 36, 44, 59, 63, 74, 90, 120, 126, 175, 177, 185, 192, 203, 209
design 24, 47, 49, 61, 64–66, 71–73, 81, 87, 112, 123, 132, 139, 144, 146, 156, 159, 162, 176, 180, 188, 191, 193, 194, 207–212, 216, 224, 248
detection 5, 30, 32, 132, 153
devices, relief 40, 56, 61, 63, 65, 66, 71–73, 76, 78, 110, 112, 118, 121, 153, 156, 174, 202, 203, 216, 224, 230
Dewar flask 58, 59, 71, 72, 84, 225
disinfectant 133–135, 141, 143
distance 29, 71, 79, 114, 151, 155, 156, 165
dose 6, 105–107, 139, 151, 152, 164–166, 249
double-gloving 108, 109
dry ice 55, 57, 61
Dufault, Michele 4–6, 44

E

Ebola virus disease 132
E. coli 122, 125, 129, 145, 147, 231
electrical equipment 75–78, 94, 149
electricity 73–76, 92, 93, 159, 215, 216
electromagnetic fields, radiofrequency 79, 153, 155, 156, 215, 216
electromagnetic radiation 79, 148, 150, 154, 155
electropolishing 37, 199
emergencies 38, 71, 216
emergency response 71, 113, 114, 117
energy 6, 13, 19, 27, 40, 52, 57, 64, 78, 79, 90, 91, 93, 96, 111–113, 136, 142, 148–150, 153,

155, 160–162, 164, 199, 201–203, 210, 215, 216, 239, 240
engineering controls 24, 30, 31, 154, 155, 160, 165, 186
engineering research 3, 13, 15
engineers 15, 154, 182, 239
environment 3, 11–13, 15, 17, 21, 24, 26, 56, 80, 123, 125, 126, 128, 129, 132, 137, 143–145, 232, 249
equipment, vacuum 3, 4, 8, 11, 17, 21, 24, 27, 30–32, 37, 41, 42, 44, 48, 53–56, 58–61, 63, 64, 66, 71–73, 75–79, 81, 84, 85, 87–89, 93–95, 98, 99, 120, 129, 133–137, 141, 145, 149, 150, 154, 155, 159, 161, 165, 172, 173, 176, 180–182, 186, 199, 207, 209, 215, 216, 223, 227, 230–232, 241, 243, 247
equipment SOPs 231
escape 54, 93, 131, 132, 138, 141, 225
estimating risk 20, 21, 23
ethics 15, 18, 175
evacuation 71, 92, 93, 113
exothermic 53, 94, 95, 112, 232, 233
experimental design 144, 206
experimental procedures 35, 162
experimental protocols 8, 36, 162, 230, 231
explosion 3, 8, 19, 37, 40, 54, 59, 60, 64, 71, 72, 84, 89, 91, 94, 96, 98, 110, 111, 113, 114, 125, 159, 199, 208, 211, 231–233
explosives 111, 112
exposure 4, 6, 17, 24, 25, 39, 56, 61, 63, 73, 79, 81, 86, 87, 89, 97, 99–102, 104–109, 111, 117–120, 124, 126, 130, 132, 141, 151–155, 157, 159, 161, 164–166, 177, 178, 203, 216, 232, 247
exposure limits 79, 81, 107, 108, 152, 155
extinguishers 92, 93, 216
eye protection 47, 61, 73, 90, 102, 124, 135, 154, 163, 232
eye protection for chemical hazards 102
eye protection for physical hazards 47
eyes 28, 29, 47, 52, 63, 76, 89, 102, 103, 155, 157, 178, 203, 232, 233
eyewear 10, 11, 17, 47, 90, 102, 103, 129, 158, 161, 162, 164, 203, 216, 245–247
– corrective 102, 103
– protective 4, 10, 11, 17, 30–32, 48, 49, 53, 58, 61, 62, 77, 98, 101, 102, 106, 108, 109, 120, 129, 132, 145, 154–156, 158, 160–162, 164–166, 172, 173, 186, 199, 216, 245–247

F
film badges 153
fingers 8, 49, 50, 58, 105, 158, 225, 240
fire 9, 13, 19, 20, 26, 32, 35, 38, 44, 52–54, 56, 65, 67, 71, 84, 91–97, 99, 104, 110, 113, 118, 124, 143, 154, 159, 172, 173, 188, 189, 193, 199, 208, 212, 216, 225
fire-retardant clothing 99, 161
fire tetrahedron 92
Fischer burners 52, 94, 210
flame 52, 54, 83, 93, 99, 211, 212
flammability 13, 67, 86, 87, 93
flammable materials 54, 91, 94–97, 99, 124, 139, 143, 216, 225, 227, 231
flammables and oxidizers 90, 97
flow 19, 61, 65, 71, 74–76, 82, 89, 132, 138–140, 145, 155, 158, 186, 203, 216, 223, 224
– current 74–76, 78, 93, 141, 185, 191, 193, 199, 216, 244
– laminar 138–140
food 3, 15, 90, 105, 143, 172, 173, 175
freezers 56, 57, 85, 136, 216
fuel 54, 83, 90–92, 95, 96, 111–113, 212, 225
fuel sources 95
furnaces 94, 154, 216

G
gases 14, 55, 58, 60, 61, 63, 66, 67, 71, 85, 87, 93, 95, 96, 102, 110, 112, 115, 132, 172, 173, 215, 224, 225
– cryogenic 55, 57, 58, 60, 62, 63, 71, 72, 96, 136
– expansion 59, 64, 65
– flammable 9, 19, 37, 54, 56, 67, 71, 91, 93–97, 99, 104, 113–115, 117, 118, 120, 124, 139, 143, 172–174, 177, 193, 208, 215, 216, 224, 225, 227, 231
– liquified pressurized 60, 61
– permanent 22, 58, 63, 71, 77, 122, 158, 216, 232, 233
– pressurized 9, 54, 60, 61, 63–66, 71, 72, 85, 132
GFCIs (ground-fault circuit interrupters) 76, 219
glass 20, 47–50, 52, 57, 59, 73, 86, 87, 89, 91, 94, 97, 99, 101, 104, 110, 124, 136, 151, 154, 165, 210, 225, 233
glassware 49, 50, 52, 73, 97, 98, 101
Globally Harmonized System 114, 115
glove boxes 132, 133

glove material 101, 102, 129
gloves 3, 4, 48–50, 53, 57, 61–63, 77, 98, 101–103, 108, 109, 123, 129, 135, 141, 145, 173, 177, 178, 199, 216, 231, 232
- chemical resistance 101
- disposable 4, 47, 49, 50, 101, 102, 109, 123, 129, 136, 146, 178
goggles 8, 30, 47, 61, 63, 90, 98, 102, 103, 108, 135, 145, 154, 161, 162, 166, 172, 173, 178, 199, 232, 233, 243
γ-rays 153
ground 59, 74, 76–78, 83, 87, 94
ground fault 76
ground-fault circuit interrupters. *See* GFCIs 76, 219
guards and interlocks for mechanical hazards 40

H

hair 4–6, 27, 44, 54, 103
hazard analysis 26–28, 171, 174, 176–180, 187, 199, 201, 206, 248
hazard analysis techniques 28
Hazard and Operability Study 26, 28, 180
hazard and risk controls 26, 28, 30, 32
hazard classes 114, 215
hazard control process 26
hazard identification 26, 36, 176, 215, 223
hazard identification techniques 26
hazardous chemicals 3, 12, 14, 44, 63, 102, 125, 139, 232
hazardous equipment 241, 243
hazardous equipment procedures 241
hazardous gases 102, 224
hazardous materials 9, 63, 71, 73, 114, 203, 230, 232, 238, 247
hazardous materials exposure 247
Hazardous Materials Information System (HMIS)–114
hazardous temperatures 52
hazardous wastes 232
hazards 11–13, 15, 17, 20, 26–29, 31, 35, 36, 39, 40, 44, 47, 59, 60, 63, 66, 67, 72–74, 77, 79–81, 83, 85, 86, 90, 91, 100–102, 105, 108, 110, 112–118, 120, 124–126, 128, 135, 141, 144, 148, 151, 154, 158, 159, 162–164, 171, 175–177, 180, 182, 187, 193, 199, 207–209, 212, 215, 216, 240, 247, 248
- agent 123, 126–128, 131, 132, 147, 215
- basic types of 13, 47, 86
- biological 13–15, 44, 48, 49, 52, 63, 73, 94, 95, 102, 125–141, 143–146, 151, 152, 154, 175, 207, 215, 216, 227, 230, 231, 247, 249
- combustion explosion 115
- corrosivity 87, 104, 232
- electrical 41, 49, 74–79, 82, 85, 87, 94, 96, 99, 149, 181, 186, 216, 245–247
- environmental 24, 80, 116, 137, 139, 140
- evaluating 36, 176, 248
- eye 10, 11, 27, 29, 32, 47, 61, 63, 73, 90, 102, 103, 124, 135, 154, 157, 158, 162, 163, 199, 201, 210, 216, 232, 233, 241, 244
- impact 29, 47, 73, 79, 102, 110, 185, 201, 203, 210, 216, 239, 240, 244
- non-beam 158
- pressure-related 72
- procedure 4, 6, 13, 26, 28, 36, 71, 98, 120, 126, 128, 162, 164, 166, 171, 180, 199, 200, 203, 212, 225, 230, 232, 239–241, 243, 244
- puncture 48–51, 91
- radiation 6, 13, 25, 40, 52, 55, 76, 78, 79, 123, 148–156, 158, 159, 162, 164–167, 215, 216
- reactive 26, 91, 102, 110–113, 247
- reproductive 90, 102, 105, 108, 119
- security 81, 132, 171
- self-polymerization 67
- trip 35, 41, 77, 240
hazard scenario 22, 23, 26, 28–31, 176
HAZOP (Hazard and Operability) 26, 28, 181
health 3, 15, 17, 24, 37, 78, 81, 101, 107, 113, 115, 127, 128, 130, 132, 141, 149, 151, 152, 155, 158, 165, 232, 247–249
health hazards 113
health physicist 151, 152, 165
heart 63, 74, 75
heat 13, 32, 40, 44, 52–56, 71, 76, 110–112, 154, 155, 199–201, 210, 216
heat guns 54
heating 52–54, 71, 79, 97, 131, 154–156, 158
heat-resistant materials 54
heat transfer 52, 53, 55, 56, 76, 154
HEPA (High Efficiency Particulate Air) 131, 138–141, 143
hierarchy of controls 30
high pressures 59, 66, 72
HMIS (Hazardous Materials Information System) 114

hood 4, 15, 25, 39, 54, 63, 66, 85, 87–89, 94–96, 108, 120, 124, 137–139, 141, 199, 216, 225, 232, 233
 – clean air 140
 – laminar flow 138, 139
hot materials 53
hot work operations 54, 216
human immunodeficiency virus (HIV)–130

I

ignite 44, 52–54, 76, 94, 96, 98, 177
ignition 19, 54, 67, 90, 92–96, 99, 159, 193, 216
ignition sources 67, 96, 216
impact goggles 47, 102
inattentional blindness 12
inattentiveness 12
incidents 3, 10, 11, 20, 25, 48, 78, 80–82, 89, 106, 113, 125, 210
infection 125–128, 144
 – laboratory-acquired 125, 127, 128
infectious material 132, 135, 136, 144
information, hazard-related 9, 11, 37, 102, 114, 116–118, 176, 199, 203, 230, 231, 248
infrared radiation 154
inhalation exposures 87, 93
inhalation hazards 108
Inherently Safer Design. See ISD xvi, 123, 144, 188, 206, 207, 209–211, 248
injury 3, 5, 6, 15, 20, 22, 24, 26, 27, 32, 49, 56, 63, 74, 75, 76, 79, 80, 89, 104, 136, 144, 157, 159, 175, 177, 185, 194, 203, 244
interlocks 40, 41, 43, 141, 160, 216
investigator, principal 3, 9, 11, 16, 17, 19, 24, 35, 36, 40, 72, 82, 86, 114, 115, 124, 125, 128, 130, 151, 155, 177, 182, 187, 188, 216, 232, 243, 244
ionization 153
ionizing radiation 148, 150–153, 166, 216
ISD (Inherently Safer Design) xvi, 207–210

J

JHA (Job Hazard Analysis) 28, 171, 176–181, 199, 201
Job Safety Analysis (JSA) 176
JSA. See Job Safety Analysis 28, 176, 248

L

lab coat 9, 50, 52, 54, 99, 108, 145, 146, 172, 173, 199

labels 57, 91, 114, 115, 232
laboratories 3, 9, 12, 14–17, 26, 44, 47, 52, 53, 58, 63, 67, 72, 81, 82, 89–93, 97, 99, 102, 104, 107–109, 121, 127, 129–135, 146, 149–151, 154, 175, 211, 225, 232, 247, 248
 – maximum-containment 132
 – microbiology 102
 – physics 3, 4, 71, 149, 150, 171, 201, 203
 – pilot-scal 72
laboratory accidents 3, 9, 15, 22, 99, 175
laboratory-acquired infection (LAI) 125
laboratory equipment 72, 135, 136, 180, 223, 227, 230
laboratory hazards 73
laboratory practices 125, 127, 129, 130, 133, 248
laboratory procedure hazards 126
laboratory refrigerators and freezers 56
laboratory safety checklist 215, 217
LAI (laboratory-acquired infection) 125
laser beam 158–160, 162, 164, 203, 210, 216, 243
laser beam hazards 159
laser classifications 158
laser diodes 156
laser exposures 161
laser hazard controls 159
laser pointers 157, 159
lasers 14, 37, 44, 47, 53, 154, 156–162, 166, 215, 243, 244, 247
handheld 153, 157, 158, 211
laser safety 157, 159, 161, 203, 216, 243–248
laser safety specialist 161
laser safety standards 157
laser wavelengths 158, 161
lathe 5, 6, 27–31, 40, 44, 85
laws 9, 37, 41, 92, 95
LD_{50} 37, 106, 107, 123
liquid, cryogenic 5, 30, 44, 47, 55, 57–63, 65, 71, 72, 79, 82, 84, 94, 96, 102, 110, 132, 134–136, 211, 212, 225, 233
liquid air 59, 84
liquid helium 5, 59, 79
liquid nitrogen 58, 59, 71, 79, 136, 225
liquid oxygen 59, 96
liver 49, 105, 106, 119, 123, 158
Lower Flammable Limit (LFL) 37, 93
lungs 86, 89, 91, 108, 149

M

machine 3–6, 27, 40, 41, 43, 44, 63, 85, 164, 165, 178, 191, 209, 239
machine shops 27, 44
magnet 79
magnetic fields 78, 79, 155
magnetic resonance imaging (MRI) 78
magnetism 73, 216
materials, energetic 3, 5, 8, 9, 12, 25, 44, 47, 49, 50, 52–54, 56–60, 63, 64, 66, 67, 71, 73, 74, 76, 79, 82, 83, 85, 89, 90, 91, 93–101, 103, 104, 108–114, 118, 120, 124, 125, 129–131, 133–143, 145, 149, 151, 154, 159, 164, 172, 173, 178, 179, 186, 188, 199, 201–203, 207–209, 215, 216, 224, 225, 227, 230–233, 238, 240, 246–248
material SOPs (Standard Operating Procedures) 231
mechanical hazards 17, 40, 216
microorganisms 145
microtome 48, 180
mind-mapping, 35, 41, 42
mitigation 24, 28–30, 32, 192–194
mitigation credit 192–194
molecules 13, 56, 65, 100, 110, 112, 118, 121
monomers 112
motors 27, 44, 78

N

National Fire Protection Association 54, 113
National Institute of Occupational Safety and Health (NIOSH) 107
needles 48–50, 86, 91, 129, 136
negligence 15, 175
neutrons 6, 149, 151
NHZ (Nominal Hazard Zone) 159–161
NMR (nuclear magnetic resonance) 4, 59, 78, 79, 226–228
nodes 181, 185, 190
NOEL (No Effect Level) 106
noise hazards 81
Nominal Hazard Zone. See NHZ 159, 222, 247
non-beam laser hazards 158
nuclear magnetic resonance (NMR) 4, 78

O

open flames 52, 54, 94, 142, 143
organisms 13, 125, 126, 128–130, 132, 136, 141, 215
overpressure 19, 64, 71, 110, 111
oxidizer baths 98
oxidizers 86, 90, 96–98, 111, 112
oxidizing acids 100, 104, 199, 216
oxygen 54, 57, 59, 60, 66, 67, 71, 81, 83, 86, 91–93, 96, 111, 112, 203, 215, 225
oxygen levels 67

P

paragraph form 231, 232, 238, 241
particle accelerators 148, 149
particles, alpha 13, 63, 148, 149, 153, 215, 227
particulate material 138
particulate radiation 148
pathogenicity 126
PELs (Permissible Exposure Limits) 107, 108
penetration time 101, 102, 109
Permissible Exposure Limits. See PELs 107
personal protective equipment. See PPE 4, 11, 17, 30–32, 53, 58, 77, 120, 145, 154, 155, 161, 165, 173, 174, 186, 199, 200, 217, 245, 246
physical hazards 13, 27, 40, 47, 79, 83, 85, 101, 110, 215, 216, 248
pictograms 115
pinch point 40–42, 178
power 20, 32, 35, 72, 74, 77, 78, 85, 133, 155–157, 159, 162, 164, 216, 243–246
PPE (Personal protective equipment) 4, 10, 24, 30, 53, 54, 62, 90, 100, 162, 192, 193, 203, 217–219, 248
press 40, 41, 83, 85, 125, 178, 188, 190
pressure 19, 40, 54, 56, 57, 59–61, 63–66, 71–73, 81, 83, 87, 105, 108–110, 112, 115, 124, 130, 134, 136, 142, 172–174, 180, 185, 189, 192, 203, 208, 210, 216, 223, 224
– working 3–5, 12, 13, 17, 44, 47, 54, 56, 57, 60, 66, 67, 71, 75, 78, 82, 87, 89, 97, 102, 103, 106, 109, 123, 125, 126, 132, 141, 142, 149, 154, 156, 160, 161, 163, 165, 172, 173, 203, 213, 216, 225, 244, 247
pressure explosions 64
pressure-injection hazards 64
pressure regulator 19, 71
pressure safety relief 61, 66, 71, 174, 203, 216, 224
pressure safety relief valves. See PSRVs 61, 65
pressurized systems 54, 64, 66
prevention 30, 32, 56, 71, 92, 125, 132, 141, 248

probability 20–25, 27, 29, 54, 102, 127, 164, 191–194
probability ratings 22, 23
procedures, operating 8, 9, 12, 17, 21, 24, 26, 28, 29, 32, 35, 36, 48, 56, 59, 63, 71, 74, 78, 118, 126–128, 131, 136, 151, 158, 160, 162, 164, 175, 180, 181, 191, 193, 200, 201, 203, 216, 230–232, 241, 243, 244, 248, 249
properties, chemical 3, 4, 8, 12, 13, 15, 19, 25, 30, 32, 37, 38, 40, 44, 47, 48, 52–55, 60, 61, 63, 65–67, 71, 72, 82, 85–92, 94–99, 101–104, 106–118, 120, 122, 124, 125, 128, 137, 139–141, 148, 154, 157–159, 162, 171, 175, 177, 180–182, 185, 187, 199, 208, 209, 215, 216, 224, 230, 232, 233, 247, 248
protection 10, 26, 29, 44, 47, 53, 54, 61–64, 73, 76, 86, 87, 89, 90, 92, 99, 101, 102, 113, 114, 117, 120, 124, 130, 135, 137–141, 144, 148, 154, 155, 161–164, 166, 191, 192, 232, 233
– environmental 24, 80, 116, 137, 139, 140
– hearing 63, 81, 135, 158
– personal 4, 11, 15, 17, 30–32, 35, 58, 61, 73, 77, 78, 82, 114, 120, 137, 153–155, 165, 172, 173, 175, 186, 199, 216, 245
protocols 8, 36, 90, 114, 129, 162, 176, 230–232, 239, 241, 243
protons 120, 148–150
PSRVs (Pressure safety relief valves) 65, 66
public 6, 15, 17, 24, 58, 71, 80, 81, 93, 106, 129, 130, 132, 152, 165, 175, 207, 227
puncture 48–51, 91
pyrophoric materials 9, 99, 120, 215

Q
qualitative risk assessments 22
quenching 79

R
radiation 6, 13, 25, 40, 52, 55, 76, 78, 79, 123, 148–156, 158, 159, 162, 164–166, 215, 216
heavy-nucleus 149
neutron 6, 149, 151, 215
non-ionizing 123, 153, 156, 215, 216
proton 148–150
ultraviolet 76, 109, 121, 122, 141, 148, 153, 161, 166
radiation dose 164–166

radiation exposure 151–153, 164, 165, 216
radiation hazards 13, 148, 151, 164, 215, 216
– non-ionizing 123, 153, 156, 215, 216
– secondary 29, 36, 37, 55, 91, 95, 98, 104, 110, 146, 148, 151, 199, 216, 231, 233
radiation protection 162
radiation safety 154, 155, 165, 216
radiation source 149, 151, 153
radioisotopes 52, 148, 166
reactions 13, 52, 53, 55, 65, 71, 72, 74, 86, 94, 95, 100, 104, 105, 110–112, 197
reactive chemicals 113
reactivity, chemical 3, 4, 6, 8, 12, 13, 15, 19, 25, 30, 32, 37, 38, 40, 44, 47, 52–55, 61, 63, 65–67, 71, 72, 85–92, 94–98, 101–104, 106–118, 120, 122, 124, 125, 128, 137, 139–141, 158, 159, 162, 171, 175, 180–182, 185, 187, 199, 208, 215, 216, 224, 230, 232, 233, 247, 248
reference books 247
refrigeration 55, 57, 58, 66, 72
refrigerators 56, 64, 78, 85, 136
regulations 15, 16, 37, 38, 52, 93, 104, 109, 209, 247
responsibility 9, 16, 17
reusable gloves, thick 101, 102, 109, 151
RF (Radiofrequency) radiation 155
risk 3, 5, 12–15, 17, 20–30, 35, 36, 47, 49, 51, 52, 54, 57, 59, 60, 71, 73, 78, 82–85, 94, 98, 100, 101, 104, 108, 115, 120, 124, 126–129, 131, 132, 134, 135, 144, 152, 162, 165, 166, 174–178, 181, 186–189, 192–194, 203, 207–210, 232, 233, 247
– acceptable 24, 54, 91, 104, 177, 181, 216, 223, 232, 233
– biological 13–15, 44, 48, 49, 52, 63, 73, 94, 95, 102, 125–141, 143–146, 151, 152, 154, 175, 207, 215, 216, 227, 230, 231, 247, 249
– catastrophic 20, 65, 71, 188, 193, 207, 211
– high 6, 12, 17, 20, 27, 30, 51, 54, 59, 62, 63, 65, 66, 71, 72, 75, 76, 78, 83, 89, 100–102, 112, 118, 123, 126, 128, 129, 132, 148–150, 152, 154–157, 164, 178, 181, 185, 187–189, 191, 192, 199, 201, 203, 208, 212, 215, 216, 232
– low 20, 27, 29, 55, 56, 62, 71, 78, 83, 89, 97, 103, 104, 110, 118, 128, 135, 151–153, 167, 187, 189, 192, 194, 210, 212, 215, 223
– rating 22–24, 77, 113, 187

– reducing 30, 36, 83, 89, 92, 151, 155, 161, 174, 177, 194, 209
– significant 5, 23, 27, 80, 82, 98, 112, 114, 115, 120, 150, 151, 154, 156, 164, 172, 194, 216
risk assessment 23, 36, 207
risk controls 23, 24
risk groups 127, 128
risk level 181, 188
risk reduction 24, 29, 174, 187, 188
risk screening 26, 27
robot 41, 43, 47, 216
routes of exposure to chemical hazards 86
run-in points 43
rupture discs 61, 65, 66

S

safety, chemical 3, 4, 6–13, 15–20, 24–32, 35–38, 40, 41, 44, 47, 52–56, 61, 63–67, 71–74, 77, 78, 81, 82, 85–92, 94–96, 98, 100–104, 106–118, 120, 122, 124, 125, 128, 130, 132, 135–141, 145, 146, 154, 155, 157–162, 164, 165, 171–176, 180–182, 185, 187, 188, 191, 192, 194, 199, 201, 203, 207–212, 215, 216, 223–225, 227, 230–233, 239–241, 243–249
safety behaviors 17
safety culture 17
Safety Data Sheets. See SDSs 18, 100, 101, 124
safety equipment 8, 27, 137
safety eyewear 11, 17, 47, 102, 164, 203, 216, 245, 246
safety glasses 10, 29, 47, 73, 102, 135, 145, 154, 173, 203, 240
– prescription 47, 102, 232
safety goggles 8, 47, 145, 154, 172, 173, 243
safety hazards 17, 35
safety information 9, 37
safety literature 36
safety shielding 232, 233
Sangji, Sheri 9, 44, 54, 120, 121
sash 87, 89, 142, 197, 227, 232, 233
scalpels 86, 129, 144
Schlenk manifold 87, 120
SDSs (Safety Data Sheets) 101, 116, 117
secondary containment 91, 98, 104, 110, 146, 216, 233
septa 57, 91, 94, 120
severity 13, 20, 22–25, 30, 42, 97, 115, 128, 192–194

severity and probability 23, 192
sharps 48–52, 84, 85, 129, 134, 172, 173, 216, 248
sharps containers 51, 85, 173, 216
sharps disposal 50, 134
shear points 42, 43, 216
shield 8, 25, 29, 30, 61, 83, 84, 87, 98, 149, 151, 189, 192, 197
shielding 148–151, 153, 232, 233
shock 25, 32, 74, 75, 76, 84, 105, 186, 200
Short-term Exposure Limit (STEL)
shrapnel 40, 64, 71, 73, 76, 79, 83, 174, 203, 233
skin 4, 49, 52, 61, 63, 66, 75, 86, 90, 91, 100, 107, 108, 144, 154, 157, 158, 178, 203, 232, 233
Slotin, Louis 6, 7, 152
solvents, flammable 9, 19, 37, 54, 56, 67, 71, 91, 93, 94, 95, 96, 97, 99, 101, 104, 108, 113–115, 117, 118, 120, 124, 139, 143, 172–174, 177, 193, 208, 215, 216, 224, 225, 227, 231
sonicators 135, 136, 141
SOPs (Standard Operating Procedures) 24, 161, 162, 220, 222, 230–232, 243, 248
sparks 54
spills 63, 91, 96, 109, 111, 117, 146, 232
splashes 47, 63, 87, 98, 102, 135, 141, 178, 233
splash goggles, chemical 3, 4, 8, 12, 13, 15, 19, 25, 30, 32, 37, 38, 40, 44, 47, 52–55, 61, 63, 65–67, 71, 72, 85–92, 94–96, 98, 101–104, 106–118, 120, 122, 124, 125, 128, 137, 139–141, 158, 159, 162, 171, 175, 178, 180–182, 185, 187, 199, 208, 215, 216, 224, 230, 232, 233, 247, 248
squeeze point 43
Standard Operating Procedures. See SOPs xi, 160, 161, 202, 204, 222, 230, 232, 234, 236, 238, 240, 242, 244, 246
standards 5, 16, 37, 38, 54, 71, 79, 113, 155, 157, 160, 172, 188, 247
statements, precautionary 115, 116
stomachers 136
Structured What-If Technique (SWIFT) 182
superconductivity 82
surface, hot 47, 52–54, 56, 62, 63, 71, 72, 74–77, 84, 87, 92–95, 98, 102, 121, 124, 138–141, 150, 154, 156, 158, 162, 163, 167, 193, 199, 210, 216, 224

surgeon 164–166
SWIFT (Structured What-If Technique) 183
Swiss Cheese model 9, 10
syntheses, chemical 3, 4, 8, 12, 13, 15, 19, 25, 30, 32, 37, 38, 40, 44, 47, 52–55, 61, 63, 65–67, 71, 72, 83, 85–92, 94–96, 98, 101–104, 106–122, 124, 125, 128, 137, 139–141, 158, 159, 162, 171, 175, 180–182, 185, 187, 199, 208, 215, 216, 224, 230, 232, 233, 247, 248
syringes 49, 86, 120, 136

T
task decomposition 176, 177
temperatures 32, 52, 53, 55–57, 62, 76, 79, 82, 83, 110, 187, 205
– cold 13, 47, 55, 56, 58, 61
– cryogenic 55, 57, 58, 60, 62, 63, 71, 72, 96, 136
– low 20, 27, 29, 55, 56, 62, 71, 78, 83, 89, 97, 103, 104, 110, 118, 128, 135, 151–153, 166, 187, 189, 192, 194, 210, 212, 215, 223
Threshold Limit Values (TLVs) 108
torches 52, 54, 94
toxic chemicals 13, 104, 215, 249
toxic effects 87, 106, 107, 249
toxicity 13, 37, 57, 87, 93, 102, 105–109, 119, 123
– chronic 104–107, 215
toxicity hazard 87
toxins, reproductive 90, 102, 105–108, 119, 125, 139, 215
training 5, 6, 24, 29–31, 59, 60, 78, 92, 99, 100, 125, 149, 162, 186, 209, 216, 230, 244, 245, 248
tunnel vision 12
two-hand control 41

U
ULPA (Ultra Low Particulate Air) filter 138–140
unacceptable risks 17, 188
Upper Flammable Limit (UFL) 93
UV (Ultraviolet) 123, 141, 153–156, 158, 164, 167, 215, 216, 227

UV lamps 141, 154
UV light hazards 141

V
vacuum 63, 64, 72, 73, 84, 94, 99, 135, 150, 215, 216
vacuum failures 73
vacuum flasks 73, 135
vacuum sources 63
vacuum system 72, 73, 135
valve, flow control 21, 58, 65, 71, 73, 174, 202, 203, 216, 223, 224
Vapor Hazard Ratio 87, 108
vapor pressure 87, 108, 124
Vladimir Likhonos 3
volatility 86, 87, 108
voltages 12, 74–76, 78

W
waste 11, 15, 27, 51, 52, 57, 91, 97, 101, 111, 123, 124, 129, 132, 134, 136, 141–143, 172, 173, 216, 231, 232, 238, 247
waste bottle 91, 97, 111
water 6, 15, 50, 53, 56, 66, 85, 87, 89, 98, 100, 110, 117, 124, 143, 145, 148, 155, 177, 179, 182, 186, 232
wavelengths 155–159, 161, 163, 164, 244
welding 52–54, 71, 154, 225
Wetterhahn, Karen 4, 36, 90, 109
What-If technique 180–182
WHO (World Health Organization) Risk Groups 128
work, hot 3, 5, 7, 11, 13, 15–17, 30, 37, 47, 49, 50, 52–54, 56, 59, 60, 62, 66, 71, 72, 75–78, 80, 81, 83, 84, 85, 87–89, 91, 92, 94, 95, 98, 99, 102, 106, 107, 109, 117, 120, 121, 123–125, 128–130, 132, 133, 135, 136, 138–147, 151, 156, 160, 161, 164–166, 175, 177, 181, 182, 185, 189, 193, 207, 209, 216, 227, 232, 233, 247, 249
World Health Organization 127, 132, 141, 155

X
X-rays 25, 150